# PHYSICAL TECHNIQUES IN THE STUDY OF
# ART, ARCHAEOLOGY AND CULTURAL HERITAGE

VOLUME 1

*Cover photograph*: The pots are part of the Egyptian Collection of the Royal Albert Memorial Museum and Art Gallery, Exeter, UK.

PHYSICAL TECHNIQUES IN THE STUDY OF

# ART, ARCHAEOLOGY AND CULTURAL HERITAGE

*Editors*

**DAVID BRADLEY**
*University of Surrey*
*Department of Physics, Guildford,*
*GU2 7XH, UK*

**DUDLEY CREAGH**
*University of Canberra*
*Faculty of Information Sciences and Engineering*
*Canberra, ACT 2600, Australia*

**VOLUME 1**

ELSEVIER

Amsterdam • Boston • Heidelberg • London • New York • Oxford
Paris • San Diego • San Francisco • Singapore • Sydney • Tokyo

ELSEVIER
Radarweg 29, PO Box 211, 1000 AE Amsterdam, The Netherlands
Linacre House, Jordan Hill, Oxford OX2 8DP, UK
The Boulevard, Langford Lane, Kidlington, Oxford OX5 1GB, UK
84 Theobald's Road, London WC1X 8RR, UK
525 B Street, Suite 1900, San Diego, CA 92101-4495, USA
30 Corporate Drive, Suite 400, Burlington, MA 01803, USA

First edition 2006



ISBN-13: 978-0-444-52131-6
ISBN-10: 0-444-52131-3

ISSN: 1871-1731

For information on all Elsevier publications
visit our website at books.elsevier.com

Printed and bound in The Netherlands

06 07 08 09 10   10 9 8 7 6 5 4 3 2 1

# Contents

# Preface

This volume is the first of a series on "Physical Techniques in the Study of Art, Archaeology and Cultural Heritage". It follows a successful earlier publication by Elsevier (*Radiation in Art and Archaeometry*), also produced by the editors of this book, Dr David Bradley (Department of Physics, University of Surrey) and Professor Dudley Creagh (Director of the Cultural Heritage Research Centre, University of Canberra).

There has been an upsurge of interest world wide in cultural heritage issues, and in particular, large organizations such as UNESCO and the European Union are active in providing funding for a very diverse range of projects in cultural heritage preservation. It is perceived that it is essential to preserve the cultural heritage of societies, both to benefit the future generations of those societies, and to inform other cultures. Also, institutions and locations of cultural heritage significance provide an impetus for the tourist industry of a country, and for many, cultural tourism contributes substantially to their national economy.

A growing need exists for the education of conservators and restorers because it is these professionals who will make decisions on how best to preserve our cultural heritage. Therefore, the primary aim of this book series is the dissemination of technical information on scientific conservation to scientific conservators, museum curators, conservation science students, and other interested people.

Scientific conservation, as a discipline, is a comparatively modern concept. For many years, interested scientists have addressed scientific problems associated with cultural heritage artefacts. But their involvement has been sporadic and driven by the needs of individual museums, rather than a personal lifetime study of issues of conservation of, for example, buildings, large functional objects, paintings, and so on.

The contributors of this book series are from both "interested scientists" and the "museum-based scientists". The authors have been selected with an eye to involving young as well as established scientists.

The author of chapter 1, Dr Jean Louis Boutaine, was Head of the Research Department of the *Centre de Recherche et de Restauration des Musées de France* at the Louvre, at his retirement. He trained initially as a physicist in the application of non-destructive analytical techniques, and has extensive experience in equipment design, and in the application of radioisotopes to the solution of scientific problems. Dr Boutaine has had the most distinguished career within the conservation science community. Since his retirement, he has been extremely active in driving the expansion of cultural heritage research within the European Community, through involvement in EU Projects and the organization of conferences; He is the EU-ARTECH Networking Activity Coordinator. This chapter is a veritable "treasure trove" of information. It discusses the use of science and technology to study aspects of the preservation of cultural heritage taken in its broadest sense: works of art, museum collections, books, manuscripts, drawings, archival documents, musical instruments, ethnographic objects, archaeological findings, natural history collections, historical buildings, industrial heritage objects and building. This chapter explains how science and technology are used to provide information which will assist us to understand how the artefacts have been created, how they have been handled (or mis-handled) since their creation, and how we can preserve them for the culture and the pleasure of future generations.

A review of the different techniques (examination, characterization, analysis) which are applied in this discipline of "conservation science" is presented. This is illustrated by many recent examples in various cultural areas. Some major national cultural heritage institutions, as well as European networks active in this area, are indicated. An important bibliography, including websites of interest, is provided.

The author of chapter 2, Professor Franco Casali, is a physicist by training and his interests include the study of scientific conservation. He has been a researcher at the ENEA (the Italian nuclear authority) and was the Director of a Research Centre with two experimental reactors. He was also an Expert of the United Nations (IAEA) for nuclear power stations. His last position at the ENEA was as Director of Physics and Scientific Calculus Division of the ENEA. Since 1985, he has been associated with "Health Physics" at the University of Bologna. Also, he is responsible for the teaching of "Archaeometry". At the University of Bologna, he leads a group of young physicists and computer science experts, who have developed advanced equipment for both micro-Computer Tomography and for large-object Computer Tomography. He has been one of the Italian representatives in the European Neutron Radiography Working Group.

This chapter commences with a description of the physical principles underlying the techniques of X-ray and neutron and digital radiography. It then proceeds to discuss the application of these techniques for the study of objects of cultural heritage significance.

Professor Tim Wess is responsible for Chapters 3 and 4 of this volume, which were co-authored by his research associates (Jennifer Hiller, in Chapter 3, and Craig Kennedy, in Chapter 4). Professor Wess holds the Chair of Biomaterials in the Biophysics Division in the School of Optometry and Vision Science at Cardiff University. His research interests include: the characterization of partially ordered biopolymers and mineralizing systems; and structural alterations of biophysical systems due to strain and /or degradation. The systems in which he is interested contain collagen, fibrillin, and cellulose (which relate, in the cultural heritage discipline, to an interest in parchment and papers). A parallel interest is in the structure of bone and artificial composite materials (which relates to his interest in historical studies of bone materials).

Chapter 3 will describe the technique of SAXS (Small-angle X-ray scattering), and show how this has been used to study alteration to structure of minerals in the bone. Preservation of intact bone mineral crystallites has been shown to relate to the endurance of amplifiable ancient DNA from archaeological and fossil bone. Moreover, the variation in bone crystallite size and habit across a two-dimensional area has been studied in modern and archaeological samples. Finally, the alteration to bone mineral during experimental heating has been investigated.

In Chapter 4, there is a description of research being undertaken on historical parchments in collaboration with Dr K. Nielsen and Rene Larsen (School of Conservation, Copenhagen, Denmark). This research involves the analysis of the deterioration of historic parchments and also the simulation of the ageing process by induced oxidative damage. (This work has been supported by the EU 5th Framework on Cultural Heritage Conservation and the National Archive for Scotland).

The author of chapter 5, Andrew Hardy, received his D.Phil. in X-ray Crystallography, from Sussex University (UK) in 1971. He began studying Middle Eastern eye cosmetics ("kohls") in the early 1990s whilst working in Oman. He has continued this work in his present position at the School of Humanities and Social Sciences, Exeter University, Political and Sociological Studies, Exeter University. The chapter summarizes and reviews the published data on the usage and composition of kohls in ancient (Pharaonic) Egypt. It also gives some information, from later time periods, on kohl usage and its "recipes". This is followed by a brief description of the experimental techniques used in his studies of past and present Egyptian kohl samples. The techniques used were: XRPD (X-ray powder diffraction), LV SEM (low vacuum scanning electron microscopy), IR (infrared spectroscopy) and the relatively new technique QEMSCAN (quantitative scanning electron microscopy). Results are given on thirty-three samples of both old and new kohls using these analytical techniques. The old samples were obtained from the Pharaonic kohl pots shown on the front cover of this book; the pots are part of the Egyptian collection of the Royal Albert Memorial Museum and Art Gallery, Exeter, UK. Finally, there is a comparison of past and present kohl compositions, concentrating on the toxicology of lead and how it is related to the particle size of the galena present. Also, there is consideration of the cultural usage of kohl, *via* information on its containers etc., in ancient and modern-day Egypt.

## Chapter 1

# The Modern Museum

Jean Louis Boutaine

*Centre de Recherche et de Restauration des Musées de France (C2RMF), Palais du Louvre, porte des Lions, 14 quai François Mitterrand, 75001 Paris, France*
*Email: jean-louis.boutaine@wanadoo.fr*

**Abstract**

At present science and technology is being used to study many aspects of the preservation of our cultural heritage taken in its broadest sense: works of art, museum collections, artefacts, books, manuscripts, drawings, archive documents, musical instruments, ethnographic objects, archaeological findings, natural history collections, historical buildings, industrial heritage objects, and buildings. This chapter tries to explain how science and technology is used so that we may better understand how the artefacts have been created, how they have been handled (or mis-handled) since their creation, and how we can better preserve them for the culture and pleasure of future generations.

A review of the different techniques (examination, characterisation, analysis) which are applied in this discipline of "conservation science" is presented. This is illustrated by many recent examples in various cultural areas. Some major national cultural heritage institutions and also European networks which are active in this area are indicated. An important bibliography, together with websites of interest, is given.

**Keywords:** Conservation science, cultural heritage, artefacts, works of art, museum collections, non-destructive testing, analysis, preventive conservation, photography, radiography, microscopy, X-ray fluorescence, ion beam analysis, spectrometric techniques, dating.

**Contents**

## 1. INTRODUCTION

As Angelo Guarino writes in his introduction to the Italian project dedicated to the *Beni Culturali*:

> "It seems worthwhile to begin with an apparently odd question: what is Cultural Heritage? The usual answer is: 'Every object of historical and artistic interest'. However such an answer is a rather limited definition: it stresses in particular our heritage in art objects like paintings, statues and historical buildings but ignores other significant matters .... A better definition is: 'Every material evidence of civilisation'."

Let us start with this definition.

Throughout the twentieth century and the beginning of the twenty-first century, museums have become important institutions not only for culture, but also for tourism, the economy, and the political self-representation of nations. Historically, there has existed an "aristocracy" of the so-called "Fine Arts" museums, and they continue to be both important and influential. But in more recent times, there has been a growth of modern and contemporary art museums, industrial heritage, ethnographic museums, "eco-museums", and the like, which are gaining recognition through public and government support. It is trivial to say that the earth is becoming an open village, but it is true that cultural heritage seems more and more shared. What represented art in ancient times, how artefacts were manufactured, how they were exchanged between peoples, when, where and how techniques appeared, prospered or disappeared are topics of increasing interest to the public. How can we better understand art objects and cultural heritage artefacts and keep them available, in as satisfactory a condition as possible for future generations is a very significant challenge.

For the examination, characterisation, and analysis of cultural heritage artefacts or art objects and their component materials, the conservation scientist needs a palette of non-destructive and non-invasive techniques, to improve understanding of their manufacture, their evolution and/or degradation during time. This understanding is necessary to give a rational basis for the restoration and conservation of objects.

Materials of all types can be encountered, for instance:

- stones, gems, ceramics, terracotta, enamels, glasses,
- wood, paper, leather, textiles, bone, ivory,
- metals (iron and alloys, copper and alloys, gold, silver, lead ...), jewellery,
- paint layers, canvas and wooden backings, pigments, oils, binding media, varnishes, glues,
- synthetic materials manufactured during the nineteenth and twentieth centuries,
- materials of the industrial heritage,
- composite materials,

and so on.

For this mammoth task, scientific conservators need to achieve mastery of many analytical tools and acquire a great depth of knowledge in diverse disciplines, and as well, to share, compare, and evaluate the results obtained by other research teams, working to different sets of protocols. This chapter intends to illustrate the kind of assistance that science and technology can provide to a better knowledge of mankind's cultural heritage and also to the establishment of rational basis for its better conservation for the future generations.

References [1–6] give significant sources relative to conservation/restoration and conservation science and, as general sources of information, Appendix 2 gives some websites of interest and Appendix 3 mentions some of the major journals in the field of conservation science.

## 2. EXAMINATION, CHARACTERISATION, ANALYSIS OF CULTURAL HERITAGE ARTEFACTS ... WHY?

The systematic application of scientific methods and studies in the field of archaeology and art had its origin in the European research community and its first manifestations as early as in the late eighteenth century with the published work by the German scientist Friedrich Klaproth, who analysed the composition of metal coins. In the early nineteenth century, the French chemist Jean-Antoine Chaptal published studies on Pompeian pigments, whilst the British scientist Humphry Davy published results from research on pigment materials in Roman archaeological finds. Others, like Michael Faraday, studied the effects of glass as protection for paintings at London's National Gallery, and the German metallurgist Ernst von Bibra wrote a compendium of metal analysis, based on a study of museum collections.

The first museum laboratory with the goal of addressing problems in the conservation of Cultural Heritage was established in 1888 by Friedrich Rathgen, when he was appointed head of a new scientific institution, the Chemical Laboratory of the Royal Museums of Berlin. This facility's primary purpose was to contribute to the understanding of the deterioration of the collection's objects and to develop treatments to stop this phenomenon.

Throughout the first half of the twentieth century, new laboratories that were established, worked by studying the collections and using this knowledge to design treatments to improve conservation and/or restoration of objects. The initial efforts concentrated on answering analytical questions as well as those about the original technology and the materials of objects and monuments. Dedicated applied studies, as well as extensive and fundamental research were then undertaken, creating the basis of the present knowledge which helps us to define and understand the aspects of elaboration and material behaviour of cultural artefacts, and thus settling the common basis of what can now be called "conservation science".

The problems to be solved can be any of those mentioned in the following sections.

### 2.1. Determination of the nature of component materials of an artefact

The problem is to analyse and, if possible, define the natural origin of gems, stones, pigments, dyes, metals, terracotta, textile fibres, ivory, wood species, etc. This information allows us to understand commercial trade links and/or cultural exchanges which may have existed during the period of the artefact's creation. For example, the characterisation of the materials of a ceramic artefact, or the analysis of the composition of alloys of metallic objects, can constitute an essential route to establishing whether the object belonged to the history of the local populations or whether it was imported from other cultures. It gives important historical information on the existence of trade routes between peoples.

### 2.2. Dating

A necessary step is to evaluate the most likely age of an artefact. This enables us to make a diagnosis about whether the objects are copies or fakes.

The first application of nuclear physics methods in archaeology dates back to the 1940s and coincides with the discovery of the possibility to make dating through the measurement of $^{14}C$ isotopic concentration present in organic materials. This discovery was the work of Willard Franck Libby, who won the Nobel Prize for chemistry (1960). His physical method allowed experts to adjust and/or revise the dating of numerous findings which were previously achieved by traditional techniques. For example, see Higham and Petchey [204] and Tuniz *et al.* [205].

### 2.3. Determination of the creative process of a material or of the artefact itself

It is important to understand how the materials in an artefact are produced, and how the artefact is produced using those materials. For example: what are the origins of the yellow, red and black pigments of parietal paintings of the Magdalenian era in the caves of the Pyrenees? How were synthetic Egyptian blue and green pigments made? What are the methods of production of the following items: "bone topazes", archaeological bronzes, artificial patinas of bronze objects, gold or silver alloys of coins and medals? What are the pigments and body materials in: Mayan terracotta, glazed ceramics from the Italian or French Renaissance? A host of other problems exists, and research has been undertaken to determine the nature of: metal pins used for drawings; pigments derived from animal, vegetal, mineral origins; synthetic pigments; glues; glasses, stained glass; enamels; threads in textiles; weaving processes for textiles; alloys used in jewellery; assembly processes of art objects, statues, musical instruments, objects belonging to the industrial cultural heritage, ethnographic objects (gluing, welding, mechanical assemblies). The list is seemingly endless since it encompasses the whole range of human activity over the millennia for which it has existed. This underscores the fact that museum curators and conservators must have an extensive and sound scientific training.

### 2.4. Evaluation of the suffered alteration processes and estimation of their importance

Environmental conditions have a significant effect on the appearance and properties of artefacts. For example, burial alters the appearance and structure of glasses, bones, and ivory; exposure to weather and atmospheric pollutants erode stained glasses; photo-oxidation and photo-degradation occurs in varnishes, dyes, pigments, organic media, glues, paper and textile components; insects and moulds can infest wood and textiles: climatic conditions can degrade stones through the action of freezing and thaw, lixiviation, attacks due to atmospheric pollution, corrosive gas, and so on.

## 2.5. Diagnosis of previous modifications or restorations

Many artefacts, particularly those of significant age, will have been altered in some way during their existence. These modifications may have been made to satisfy modesty requirements for a particular historical time (renaissance paintings), as graffiti or overlying inscriptions (for example, Portuguese inscriptions on tables recording prior Chinese presence in the Congo (1421) [10]), and so on. It is necessary to determine what could have been functional modifications, dismemberment, and restoration practices in previous times. As well, identification of metallic inserts in statues, evidence of later repainting, lining or transposition of easel paintings, the application of protective varnishes on paintings or statues is essential before appropriate remedial action can be taken by the conservator.

## 2.6. Assistance to the conservator/restorer

The conservator/restorer must determine the alteration level of an artefact. And he must determine the compatibility between the materials and processes to be applied and the artefact and its components which are to be restored. The conservator must quickly formulate a conservation strategy for preventive conservation, and apply all necessary controls before, during, and at the end of the process of restoration.

## 2.7. Forecasting and optimisation of the short- and long-term destiny in the present conservation conditions (a discipline which is called preventive conservation)

Preventive conservation studies the compatibility of the artefacts with the architectural structure and air conditioning of museums, temporary exhibition galleries, historical buildings, libraries, archives rooms, storage areas, and transport containers. Because artefacts (usually very valuable ones), are transported between museums, and between museums and their storage facilities, the role of the transport container is not insignificant in determining the long-term well-being of the artefact.

Studies of the influence of such parameters as temperature, relative humidity, natural or artificial lighting (especially ultraviolet radiation), corrosive gas, dust, bio-deterioration, pollution generated by the public, vibration etc. on the durability of the artefacts must be undertaken to optimise their environmental conditions, and enhance their well-being.

Studies on the compatibility of newly produced materials, potentially usable for restoration, with the artefacts (varnishes, glues…) are being conducted. Can, for example, modern engine oils be used in old engines?

The discipline of preventive conservation must be given greater prominence in the administration of museums, libraries, and galleries in the next decade. Since the concept of national cultural heritage stems from a notion of national identity, political authorities must become more strongly involved, promoting the conservation of the past in accordance with the concept of sustainable development for the future.

A basic bibliography on preventive conservation is given in the Refs. [7–26].

## 3. INSTITUTIONS AND NETWORKS ACTIVE AT THE INTERFACE BETWEEN "SCIENCE AND TECHNOLOGY" AND "CULTURAL HERITAGE"

### 3.1. National institutions

According to various parameters relevant to national traditions and political structures, centralised or decentralised state, relative weight of the public service, relative weight of private foundations, different types of institutions or structures can play a permanent and significant part at the interface between "Science and Technology" and "Cultural Heritage": in other words, in the discipline of "Conservation Science". These institutions can be national and/or provincial cultural heritage institutions, museums, libraries, or archives with their own laboratories or scientific departments, universities or higher education establishments, restoration workshops having some Research and Development (R & D) capabilities, private and/or industrial foundations, industrial technology research centres, R & D laboratories of industrial companies active in materials used in the cultural heritage area (paper, leather, wood, pigment, dye, glass, mortar, stone, ceramics, textile …).

Appendix 1 gives a short list of some major national cultural heritage institutions in a number of countries.

### 3.2. National networks

In order to better use the knowledge existing in such various structures, to improve human and technical potential, and to share knowledge, some national institutions have taken the initiative to create dedicated networks or co-ordinated research programmes. Here, are given some significant examples at the interface between "Science and technology" and "Cultural Heritage".

#### *3.2.1. Progetto finalizzato Beni Culturali*

This important project was established by the CNR (Consiglio Nazionale delle Ricerche) in Italy, on the Safeguarding of Cultural Heritage and was started in January 1996 to continue for five years.

The Project was divided into five subprojects, four of them concerning cultural heritage artefacts:

Subproject 1:

- Archaeology and Geographical Information Systems (GIS) which are necessary to safeguard ancient resources constantly in danger of environmental and human aggression.

Subproject 2:

- Development of new scientific and technological methodologies for researches on the state of conservation of art objects.
- Development of new materials and procedures to restore and save these “art objects”.
- Development of new technical and legal procedures to prevent the impoverishment of Cultural Heritage of the Nation.

Subproject 3:

- Studies on paper decay under the action of biological and physico-chemical agents.
- Studies on new materials and procedures to restore damaged books and archive documents.
- Studies on restoration of photographic plates, films, and computer magnetic tapes.

Subproject 5:

- Innovative methodologies devoted to a better organisation and management of different typologies of museums.
- Restoration and exhibition of scientific and musical instruments.
- Exploitation of multimedia technologies with reference to different typologies of museums.

Visit http://www.pfbeniculturali.it/index01.asp for more detalis.

#### *3.2.2. ChimArt*

*ChimArt* is a “Groupement de Recherche” (GdR) of the CNRS, a grouping together of 23 French laboratories (from the Ministry of Culture, CNRS, CEA, Universities, regional restoration workshops). This network has been in existence for four years, starting January 2000, and has been further renewed for four more years. Three items have been given prominence:

- understanding of the physico-chemical mechanisms of elaboration of cultural heritage artefact materials;
- understanding of the physico-chemical mechanisms which drive the alteration processes of these materials;
- study of products used for restoration and conservation of cultural heritage artefacts and their potential interaction with the artefact materials.

Visit http://www.c2rmf.fr/homes/liens_gdr.htm for more details.

### 3.3. European networks

For conservation scientists, the evidence and the usefulness of working in the frame of European research networks has been established. The similarity of problems to be solved, the complementary nature of certain teams, the need to consolidate practices and in the near future, the need to establish European standards in the area of cultural heritage, were and will remain important as will shared motives. It is important to note that a new technical committee of the European Committee for Standardisation (CEN), dedicated to the "Conservation of Cultural Property" (CEN/TC 346) had its inaugural meeting in June 2004.

Visit http://www.cenorm.be/CENORM/BusinessDomains/TechnicalCommitteesWorkshops/CENTechnicalCommittees/TCStruc.asp?param=411453&title=CEN%2FTC+346 for more details.

#### *3.3.1. COST G1*

COST G1 was a research network, devoted to ion beam analysis applied to art and archaeology, active between 1995 and 2000. A final report has been published [27].

Visit http://www.uia.ac.be/u/costg1/home.html for more details.

#### *3.3.2. COST G7*

COST G7 is a research network dedicated to "Artwork Conservation by Laser". It has been set up to address challenges in three main areas:

1. laser systems for investigation and diagnosis,
2. laser systems for real-time monitoring of environmental pollution,
3. laser systems for cleaning applications.

A very important contribution of this COST Action is the prevention of cultural heritage deterioration. Development of techniques for monitoring the quality of indoor and outdoor atmospheres is proposed in parallel with restoration and conservation work.

Visit http://alpha1.infim.ro/cost for more details.

#### *3.3.3. COST G8*

COST G8 is a research network, devoted to the non-destructive analysis and testing of museum objects. This network, grouping together representatives from 21 countries started in December 2000 and was active till August 2005 [28].

Visit http://www.srs.dl.ac.uk/arch/cost-g8 for more details.

#### *3.3.4. ENCoRE*

ENCoRE was founded in 1997 with the main objective of promoting research and education in the field of cultural heritage, based on the directions and recommendations given in the Professional Guidelines of the European Confederation of Conservator–Restorers Organisation (ECCO) and the Document of Pavia of October 1997. Currently ENCoRE has 30 full members and four associate members from amongst the leading conservation–restoration study programmes in Europe. In addition, 21 institutions and organisations working in the field of cultural heritage protection and research are partners of the network.

Visit http://www.encore-edu.org/encore for more details.

#### *3.3.5. LabS TECH*

LabS TECH [29] is a European research network, devoted to the sharing and the enhancement of examination, characterisation, analysis, restoration and conservation methods of cultural heritage artefacts in the European Countries. The nucleus of this network comprises representatives from seven European countries (Belgium, France, Germany, Greece, Italy, Portugal and United Kingdom) plus ICCROM and USA. It was started in January 2001. It is open to cultural heritage institutions, museums, libraries, universities, research establishments, non-profit foundations, restoration workshops, industry co-operative technical centres, and private industry research laboratories active in these fields. At present, 116 institutions from 26 countries have volunteered to collaborate with the network.

The main characteristics of these institutions, together with a database on the techniques used and the cultural areas in which they are working are mentioned in the website http://www.chim.unipg.it/chimgen/LabS TECH.html.

Several open international workshops were organised on different themes: binding media identification in art objects [30], painting technique of Pietro Vannucci called "il Perugino" [31], silicon-based products in the sphere of cultural heritage [32], and novel technologies for digital preservation information processing and access to cultural heritage collections [33].

### *3.3.6. EU-ARTECH*

Following LabS TECH, a new project called EU-ARTECH (Access Research and Technology for the Conservation of the European Cultural Heritage) has just commenced (1 June, 2004) for a duration of five years, within the 6th European Framework Programme, as an Integrated Infrastructures Initiative, which includes Networking Activities, Joint Research Activities and Transnational Access to scientific instrumentation.

The ACCESS activity consists in two different noticeable opportunities open to users working in Europe and associated countries:

- AGLAE, located in the C2RMF, where non-destructive elemental ion-beam analyses (IBA) are carried out with high sensitivity and precision, for 230 person*days available during the five years of the project.
- MOLAB, a unique collection of 10 portable instruments, together with competences on methods and materials, operated by a unified group of 4 Italian laboratories, allows performing *in-situ* non-destructive measurements for studies on artworks and for the evaluation of conservation–restoration methods, directly in a museum room, or on the scaffolding of a restoration workshop, or at an archaeological site (220 person*days available). The first MOLAB measurement campaign took place in the Musée des Beaux-Arts & d'Archéologie de Besançon (France) to make a systematic survey of the paintings "*Lamentation over the dead Christ*" by Agnolo Bronzino, before an important restoration work.

Thirteen institutions from eight European countries (Belgium, France, Germany, Greece, Italy, Netherlands, Portugal and United Kingdom) participate in this project.

Visit http://www.eu-artech.org for more details.

Two first International workshops have already been organised by EU-ARTECH:

- Raphael's painting technique: working practices before Rome – London – National Gallery – 11 November, 2004 [34];

- Non-destructive analysis of cultural heritage artefacts – in co-operation with COST G8 – Amsterdam – ICN – 12 January, 2005.

In Appendix 3, one can find also information relative to other networks, working in similar areas.

## 4. MAIN TECHNIQUES USED IN THE STUDY OF CULTURAL HERITAGE ARTEFACTS

### 4.1. Specific situation of cultural heritage examination and analysis

Due to the broad diversity of materials, and as the artefacts have often various complex and undetermined compositions their elaboration processes often unknown or at least uncertain, it is generally useful or necessary to combine various examination, characterisation, and analysis methods, in order to get pertinent information (please consult the recent books published by Ciliberto [35], Creagh and Bradley [36], or Janssens [37] that cover a wide spectrum of details, or those dedicated to particular types of materials [38–40]).

Furthermore, because of the unique or rare nature of cultural heritage artefacts, as a general rule, the techniques which can be used must be either well tried and proven non-destructive and non-contact methods without any sampling, or be tests with strictly authorised small-size sampling. Table 1 indicates the most mentioned techniques presently

**Table 1.** LabS TECH – Frequency of use of the different techniques (January 1, 2005). N.B. 114 different techniques are indicated by the 116 participants

| Rank | Technique | Number of times mentioned |
|---|---|---|
| 1 | Reflection Light Microscopy | 87 |
| 2 | Scanning Electron Microscopy (SEM) | 76 |
| 3 | Transmission Light Microscopy | 75 |
| 4 | Classical Visible Light Digital Photography | 70 |
| 5 | Classical Visible Light Silver Emulsion Photography | 67 |
| 6 | Infrared Spectrometry | 60 |
| 7 | Powder Diffractometry | 49 |
| 8 | Diffractometry | 48 |
| 9 | Ultraviolet Fluorescence Photography | 47 |
| 10 | Visible and Ultraviolet Spectrometry | 47 |
| 11 | Standard Colorimetry | 47 |
| 12 | Digitisation and Image Archiving | 44 |
| 13 | Infrared Spectrometry Microscopy | 43 |
| 14 | Low HV (<150 kV) X-ray Radiography | 42 |
| 15 | Environmental Weathering Tests (Chambers) | 42 |

*Continued*

**Table 1.** Continued

| Rank | Technique | Number of times mentioned |
|---|---|---|
| 16 | Gas Chromatography (GC) | 38 |
| 17 | High Performance Liquid Chromatography (HPLC) | 38 |
| 18 | Gas Chromatography–Mass Spectrometry (GC-MS) | 36 |
| 19 | Low Angled Photography | 33 |
| 20 | Differential Thermal Analysis (DTA/TG/DTG) | 33 |
| 21 | Infrared Reflectography using an Electronic Camera | 32 |
| 22 | Infrared Silver Emulsion Photography | 31 |
| 23 | Universal Mechanical Testing | 31 |
| 24 | X-ray Fluorescence Analysis – X-ray Tube – Laboratory Fixed Instrument | 30 |
| 25 | Spectro-Photo-Colorimetry | 29 |
| 26 | High voltage (150 < HV < 450 kV) X-ray Radiography | 28 |
| 27 | Accurate Colour High Resolution Digital Photography | 28 |
| 28 | Ion Chromatography | 28 |
| 29 | Raman Spectrometry | 25 |
| 30 | Thin layer Chromatography (TLC) | 24 |
| 31 | Electron Microprobe | 24 |
| 32 | Atomic Absorption Analysis (AAA) | 23 |
| 33 | X-ray Fluorescence Analysis – X-ray Tube – Portable | 23 |
| 34 | Atomic Emission Spectrometry (ICP-AES) | 23 |
| 35 | Pyrolysis Gas Chromatography (Py-GC) | 19 |
| 36 | Environmental Natural Weathering Tests (Outdoor) | 19 |
| 37 | Mercury Porosimetry | 18 |
| 38 | Particle Induced X-ray Emission (PIXE) | 17 |
| 39 | Pyrolysis Gas Chromatography Mass Spectroscopy (Py-GC-MS) | 16 |
| 40 | Inductively Coupled Plasma Mass Spectrometry (ICP-MS) | 16 |
| 41 | Electron Impact Mass Spectrometry (EI-MS) | 15 |
| 42 | Ultra-Sound Testing | 13 |
| 43 | Contact Angle measurement | 13 |
| 44 | Specific Surface Area Measurement (BET) | 13 |
| 45 | Fluorescence Spectrophotometry | 13 |
| 46 | Rutherford Backscattering Spectrometry (RBS) | 13 |
| 47 | Environmental Scanning Electron Microscopy (ESEM) | 13 |
| 48 | Synchrotron radiation examination | 13 |
| 49 | Scanning Infrared Reflectometry | 12 |
| 50 | Thermoluminescence Dating (TL) | 12 |
| 51 | Transmission Electron Microscopy (TEM) | 12 |
| 52 | Laser Ablation Mass Spectrometry | 11 |
| 53 | X-Ray Induced Photoelectron Spectrometry (XPS) | 10 |
| 54 | Ultraviolet Fluorescence Microscopy | 10 |
| 55 | Nuclear Reactions (PIGE-PIGME) | 10 |
| 56 | Environmental monitoring | 10 |

operated by the participants of the LabS TECH network, a list which illustrates the large palette of techniques actually used.

As well, within the LabS TECH network, a questionnaire has been sent to the several participating institutions, to explore potential medium-term development of examination and analysis techniques dedicated to cultural heritage materials. This survey was conducted at the end of year 2003 and the beginning of year 2004. As an indication of prospective and future development, Table 2 gives the more frequently mentioned techniques reported by the 22 participating institutions who replied to the questionnaire.

One can make the following comments:

- Infrared spectrometry (already used by 50% of the participants) will see increased application, particularly in the near infrared range and/or through the introduction of fibre optics components in the instrumentation. The advent of synchrotron radiation IR will further enhance the usefulness of this technique for those samples which can be transported to synchrotron radiation sources. Please see http://srs.dl.ac.uk/arch/index.htm
- The Raman spectrometry technique (which is only presently used by 20% of the participants), is likely to become more widely used, both quantitatively and qualitatively ("micro Raman" and/or portable instrumentation).
- Portable energy-dispersive X-ray fluorescence technique seems to benefit by new tools like micro capillary X-ray optics (*cf.* the various contributions to the recent EXRS 2004 conference in Alghero [41]).
- Important efforts are on or will be made for rendering instruments portable for on-site measurements, as a large proportion of cultural heritage artefacts are non-movable, or are generating safety issues when being moved to examination laboratories.
- Many teams are working on the question of dual- or multitools (XRF/XRD, Raman/IR, Raman/XRF, multispectral scanning or mapping instrumentation).
- Surprisingly, very little effort appears to be put in the R & D segment concerning environmental monitors, an area which is certainly of great significance, both for the long-term conservation of artefacts in large cities and from an economic point of view, even if this probably results in less "nice publications".

**Table 2.** LabS TECH survey – Medium term development prospective (among 22 answers)

| Analytical technique | Frequency |
|---|---|
| IR Spectrometry (including FT or fibre optics) | 7 |
| Raman Spectrometry (including μRaman) | 6 |
| XRF (mainly portable or μXRF) | 5 |
| Optical microscopy–transmission | 4 |
| Standard colorimetry | 4 |
| Infrared spectro-microscopy | 4 |
| Visible and ultraviolet spectrometry | 4 |
| Laboratory environmental weathering (in climatic chambers) | 4 |

## 4.2. Examination techniques

### *4.2.1. Visual examination*

The expert's eye, eventually assisted by a magnifying glass or a binocular device, remains of course the indisputable tool for the first step of the examination process.

### *4.2.2. Photography*

This is the most used technique in scientific conservation.

For instance, for each easel painting studied, the typical sequence of examination is as follows:

- conventional reflection visible light photography, colour or black and white;
- low-angled light photography;
- reflection infrared photography ($\lambda = 750$–$900$ nm). For example, see Mairinger [42];
- ultraviolet fluorescence photography ($\lambda = 320$–$400$ nm);
- infrared reflectography using an electronic camera ($\lambda = 1800$–$2500$ nm).

Recently (since December 2003), a new development relative to a digital multispectral photography protocol occurred at the C2RMF [43]. The equipment and the protocol adopted permits one to realise sequentially, with the same operating conditions: classical photography, infrared photography, UV fluorescence photography, and raking light photography. The equipment consists in a Hasselblad still digital camera, H1 type, auto-focus with adapted lens ($F = 80$ mm), Imacon CCD detector $4000 \times 5000$ pixels, 8 or 16 bits, equivalent sensitivity 50 ISO (in practice, can be operated up to 200 ISO), useful wavelength $\lambda \leq 1050$ nm (N.B. for silver halide films, $\lambda \leq 900$–$1000$ nm), used with a video monitor.

Examples of the application of this instrumentation are wall paintings of the Galerie d'Apollon, Musée du Louvre (Paris), *Triomphe de Cybèle & Triomphe des Eaux* by Joseph Guichard, before restoration. In this case, the sketch was 12 m in length, and distance from object to camera was 25 m. For UV fluorescence, one can use a classical Broncolor flashlight without protective cache, with 3–5 flashes. For IR photography, one can use a Wratten 89 filter transparent to infrared, the sensor being modified on C2RMF request, with the infrared-absorbing filter being dismounted, and set on demand, outside the camera.

### *4.2.3. Optical microscopy*

Different types of optical microscopes are routinely used:

- reflection metallography microscope for polished samples,
- transmission petrography microscopes for thin layers ($t = 30$ μm).

### *4.2.4. Scanning electron microscopy and associated X-ray spectrometry analysis*

Scanning electron microscope (SEM) is one of the more frequently used equipment (magnifications of 200–10 000). Associated with this machine, are equipment for microanalysis using ($e^-$, X) fluorescence with the following characteristics:

- analysis of samples,
- $Z > 6$–$8$ (C to O),

- lower limit of detection approximately $10^{-6}$,
- surface examination spot diameter from some nanometres for the image to some micrometres for the analysis, associated with concentration mapping software.

Some examples using SEM are: the examination and analysis of painting materials in cross sections, the determination of multi-component composition profile in the corroded superficial layer of archaeological alloys, the analysis of mineral components of terracotta, pottery, enamels, ceramics, rocks, gems, pigments, mineralised wood, or textiles [44,45].

#### *4.2.5. Radiography [46–53]*

This technique uses non-destructive examination by transmission of a penetrating ionising radiation through the object to be controlled. The radiation is emitted by a source and detected by an appropriate detector, generally a silver halide emulsion. Various interaction phenomena can occur in competition with one another: true absorption, diffusion, emission of secondary radiation. This will differ in probability according to the nature of the radiation, its energy, the nature of the constitutive materials of the object …

If one represents this interaction phenomena by Beer's law ($I = I_0 \, e^{-\mu x}$), the linear attenuation coefficient $\mu$ will be the decisive parameter. Thus,

- For X-ray photons, specially those of low energy ($E < 100$ keV), high sensitivity to the atomic number of the examined material (1 mg cm$^{-2}$ of lead will absorb more than 1 mg cm$^{-2}$ of iron, or 1 mg cm$^{-2}$ of aluminium and *a fortiori* than 1 mg cm$^{-2}$ of an organic or plastic material);
- For electrons, near insensitivity to the atomic number (1 mg cm$^{-2}$ of paper will absorb the same as 1 mg cm$^{-2}$ of a metal).
- For thermal neutrons, interactions at the nucleus level and great dispersion of the attenuation coefficient values (Gd, H, B, Sm, Co, Eu, Li , Dy, In, Hg, and so on will absorb much more than Pb, U, Bi, Ba, Ga, Sb, Pd …)

So, the task of the operator consists in adapting these different interaction modes to the examination question relative to the artefact. The operator may have to use the following techniques.

- Low energy X-ray radiography (HV = 15 to 60 kV), large source to object distance, low speed high definition film (including large size ones up to 40 × 150 cm), similar to medical radiography or to industrial radiography as used in aviation industry. Such a technique has been used for a long time [54–65], in particular for the examination of easel paintings. In some cases, it can be pertinent to use filtered radiation in order to take benefit of singular K edge absorption discontinuities [66].
- High energy X-ray radiography (HV up to 450 kV) can be used for the examination of objects like stone or bronze statues, furniture, jewellery, pottery, ceramics, musical instruments, and so on [67–70]. The technique is very similar to industrial radiography in the foundry industry i.e. the usage of industrial radiography films with lead intensifying screens, or radioscopy devices. For very large objects, use of the facilities for X-ray examination afforded at customs facilities or aerospace industries could be considered. Energies up to 6 MV are available in both single and dual view.

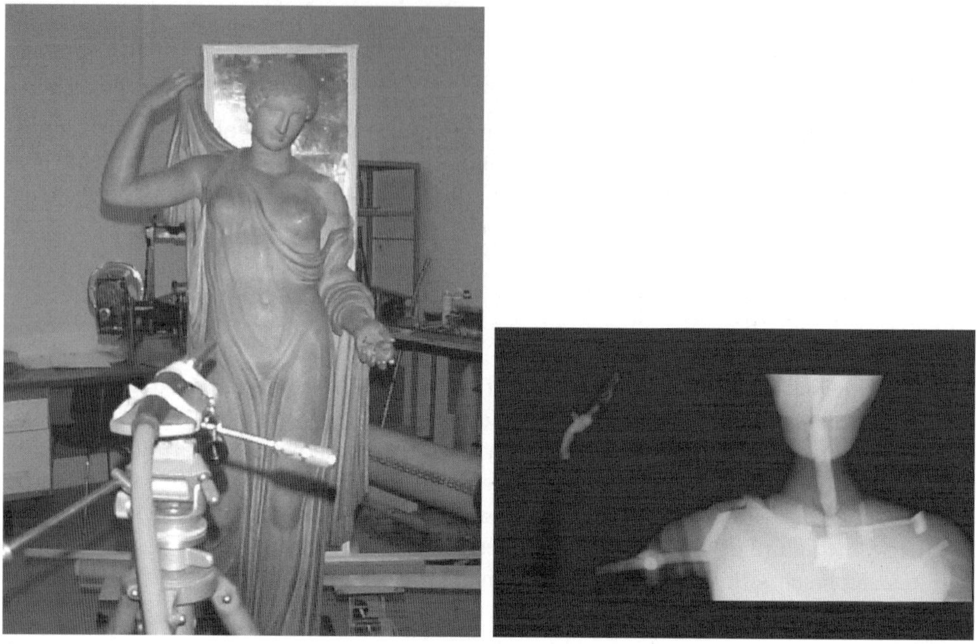

A B

**Fig. 1.** Venus Genitrix (Louvre Museum) – (A) Photograph showing the $\gamma$-radiography setup; (B) Gamma radiograph ($^{60}Co$) of the statue showing repairs (B. Rattoni, CEA).

- Beta radiography is dedicated to examination of thin foils, mainly paper, using a plane sheet source of $^{14}C$ radiolabelled plastics (poly-methylmethacrylate). ($E_{\beta\,max} = 156$ keV; $T = 5730$ years), and a fast monolayer film, used for industrial radiography or graphic arts. This permits one to accurately determine the paper structure and specially, its watermark [71,72].
- Electron emission radiograph: An X-ray generator (HV set to about 300 kV, high filtration (10 mm Cu), monolayer radiographic film in direct contact with the examined surface), is placed towards the incident beam. The surface layer of the object acts as a photon/electron converter. This technique is used for the examination of paint layer on canvas or wood backings, or enamel on copper alloy substrate [73–75]. Figure 2 shows a Champlevé enamelled object and the classical X-ray and emission radiographs taken from the object.
- Laminography: The X-ray source and the detector are moved synchronously, in order to get a sharp image only on a particular stratum of the object (used for paintings on a wood backing).
- Gamma radiography: A projector equipped with an $^{192}Ir$ source for up to 300 mm of stone or $^{60}Co$ source for up to 450 mm of stone, is used for the examination of large thickness statues (for example: marble – metopes of Olympia, Borghese gladiator; sandstone – Khmer statues from Angkor) [76–78]. Figure 1(A) shows a statue of Venus, and Fig. 1(B) shows the location of repairs which have been made.

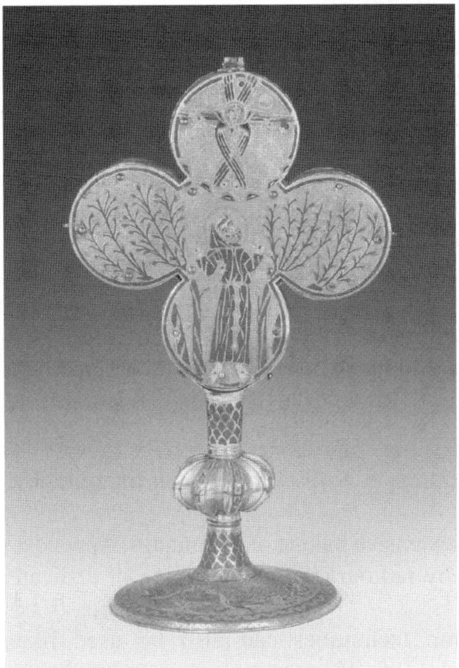

A

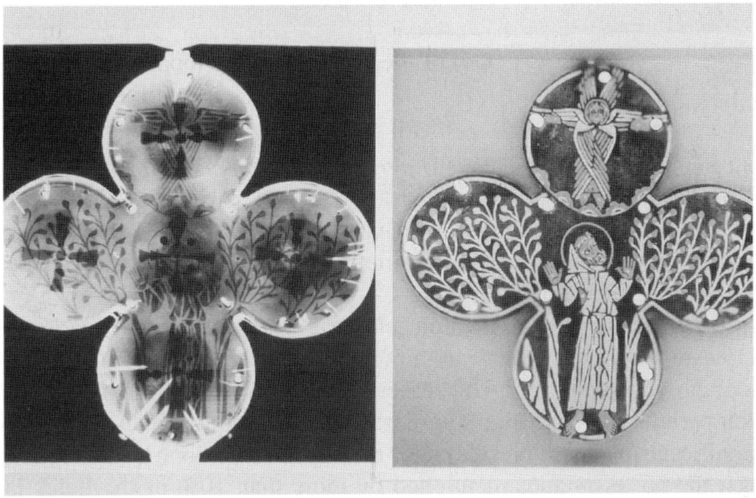

B

**Fig. 2.** Episcopal Champlevé enamelled cross from Limoges (Musée de Cluny – Paris) – (A) Photograph; (B) X-ray radiograph (left) and electron emission radiograph (right) of the object (T. Borel and D. Bagault, C2RMF).

- Neutron radiography and autoradiography: Two different methods can be operated on an extracted thermal beam of a nuclear research reactor:
  - Neutron radiography, based on the variations of mass attenuation coefficients, quite different from those of X or $\gamma$ photons, or from electrons or $\beta$ particles, is indeed scarcely used [79,80].
  - Autoradiography, obtained by activation of certain components, after neutron irradiation in a beam similar to the one used for neutron radiography is applied in some places. The film is similar to the one used for beta radiography [81–84].
- Tomodensimetry: This technique is occasionally used. Sometimes, one uses a medical scanner (X photons of low energy ($HV < 150$ kV), eventually industrial high energy machines designed for the space industry (HV up to some MV). Examples include: funeral objects made of a lead nucleus in a copper cover engraved with animal designs made of shells, from the Makran civilisation (Pakistan), and Egyptian mummies.

The reader should refer to the chapter on X-ray and neutron tomography by Casali in this book (Chapter 2).

As an illustration of the use of this extensive palette of techniques, Appendix 4 gives some examples of questions to be solved by radiography and the type of dedicated techniques applied for this purpose.

Other non-destructive examination techniques can also be used like ultra-sound, holography, and infrared thermography [85–87].

### 4.3. Analytical techniques

Two recent conferences were totally ("*Materials aspects of art characterisation, conservation and restoration*", Strasbourg, France, 2004 [88]) or partially ("*EXRS 2004 European conference on X-ray spectrometry*", Alghero, Italy, 2004 [41]) dedicated to analytical techniques applied to cultural heritage artefacts. The presented papers illustrate clearly the present tendencies in this discipline.

Amongst the palette of examination, characterisation, and analysis techniques, those based on ionising radiation occupy an important place (Wilhelm Röntgen himself made the first radiograph of a painting!). This arises from the penetrating power of ionising radiation, which permits one to design non-destructive and non-contact tools, giving information on significant thickness or volume of the object to be controlled.

Amongst the 56 techniques mentioned by more than 10% of the LabS TECH participants, 17 techniques are based on the use of ionising radiations, and 10 are based on X-ray spectroscopy techniques.

As another indicator, if one accesses the Wiley–Interscience website, based on the request "(analysis or examination or characterisation) and ('cultural heritage' or 'art object' or 'archaeological object' or archaeometry)", among the 74 pertinent references between 1997 and 2004 obtained, 24 deal with X-ray spectrometry techniques.

A recent special issue of *X-ray Spectrometry* was dedicated to the analysis of cultural heritage materials [89].

*4.3.1. X-ray fluorescence analysis*

The importance of X-ray fluorescence analysis in the area of cultural heritage is well established. Recently, progress in X-ray tube technology, capillary X-ray optics design, room temperature or Peltier effect cooled detectors, and miniaturised electronics have made possible the design of compact and transportable analysis equipment, to be used on site for archaeological excavations, historical buildings, museums, restoration workshops, and so on.

On the European CORDIS website, one can observe that many projects were recently dedicated to the design of various configurations of X-ray fluorescence devices for cultural artefacts analysis purpose (teams from Aarhus, Antwerp, Berlin, Milano, Niewegein, Sassari ...). Like other laboratories, C2RMF has developed a prototype of such equipment.

Total-reflection X-ray fluorescence analysis was recently added to the palette of techniques [90–92].

*4.3.2. Ion beam analysis (IBA) [93–98]*

COST G1 was the place for making the assessment of the possibility of the analytical techniques based on ion beams [27]. Among the participants, C2RMF is the only laboratory in the world, equipped with its own accelerator dedicated to cultural heritage artefacts analysis. AGLAE (Accélérateur Grand Louvre pour l'Analyse Elémentaire) is a National Electrostatics Corp. tandem accelerator (2 MV). Figure 3 shows the end-stations used for analyses at AGLAE.

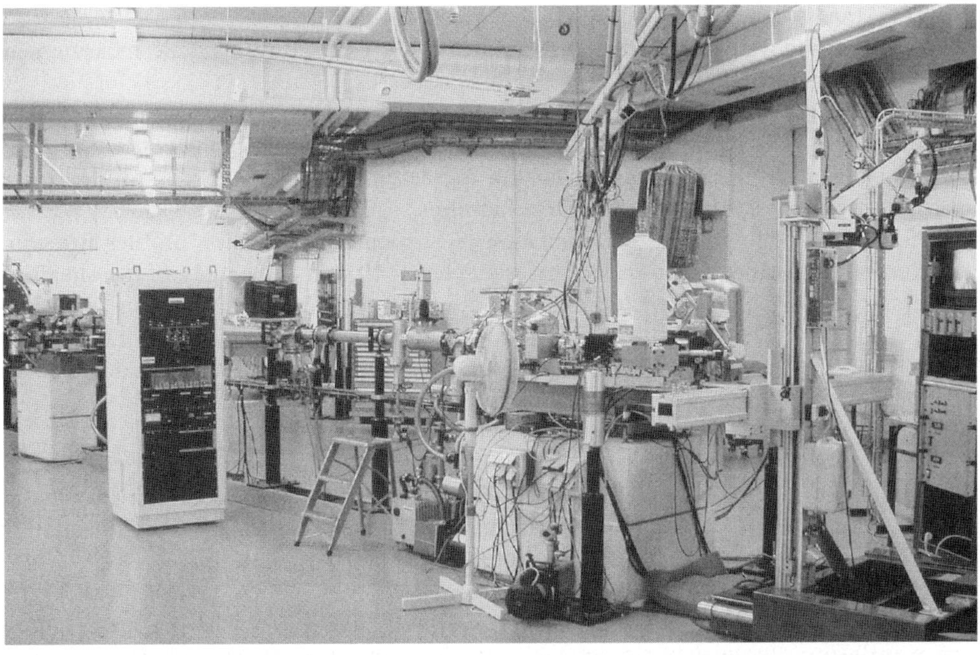

**Fig. 3.** View of the end-station of the particle accelerator AGLAE (C2RMF – Paris), showing the beam line and the solid state detectors used for IBA analysis.

Recently, a special session dedicated to accelerator applications for the analysis of art materials was organised during the ECAART-8 2004 conference [93].

The palette of IBA techniques covers the following techniques.

4.3.2.1. PIXE. The more frequent analysis mode is PIXE (Particle Induced X-ray Emission). The main characteristics are:

- analysis can be performed directly on the object, but restricted to the surface of the object, with an open air beam facility, through a very thin window (0.1 μm $Si_3N_4$), and a helium flux (no sampling) [97,98];
- $Z > 9$ to 11 (O to Na);
- practical minimum detection limit: approximately $10^{-9}$, thus possibilities of trace detection and analysis;
- approximate beam spot diameter: 10 μm to 1 mm on the surface of the object, and step-by-step scanning device.

Recent applications involved establishing the geographical origin (Burmese) of inlay rubies of the alabaster Parthe Ishtar statue (~200 BC) held in the custody of the Louvre Museum [99] (Fig. 4(A–C)), mapping of inclusions in gemstones [100,101], of pigments in illuminated manuscripts [102–104], of metal pin Renaissance drawings by A. Dürer, Pisanello, and others [105–107], characterisation of ancient metallurgical processes [108,109], study of the glazing technique of Renaissance terracotta statues [110–112] or lustre ceramics [113–115], study of the lixiviation process of buried lead glasses [116,117], and so on.

A quite new development concerns the possibility of making dynamic measurement during physico-chemical processes on solids or aqueous solutions [118].

4.3.2.2. RBS (Rutherford Backscattering). This method determines the concentration of various elements at the surface layer and/or measurement of the thickness of this layer, from the energy spectrum of backscattered protons.

It finds applications in the determination of element concentration profiles in patina layer on metallic objects (bronze, silver, lead, etc.) [119–121], the study of the alteration processes of lead objects (papal bulls), and the control of their conservation conditions using lead reference samples (Fig. 5) [122–124]. Also, this technique is one amongst others for the study of the corrosivity of atmosphere in museums, historical buildings, archives, repositories…, using metal foils and/or various sensors [125–127].

4.3.2.3. Nuclear reactions. One uses specific nuclear reactions, generally threshold ones, as (p,n), (p,2n), (d,n), etc. to determine lightweight element concentration in metallic matrixes. An example of the use of this technique is the non-destructive determination of oxygen content in archaeological bronze objects [119].

4.3.2.4. Secondary X-ray fluorescence, called $(PIXE)^2$ [128]. To achieve X-ray fluorescence analysis of lightweight elements in a matrix of heavyweight elements, a solution can be found using a beam line equipped with an intermediate target acting as a proton-induced low-energy secondary X-ray source of an element with an atomic number between those

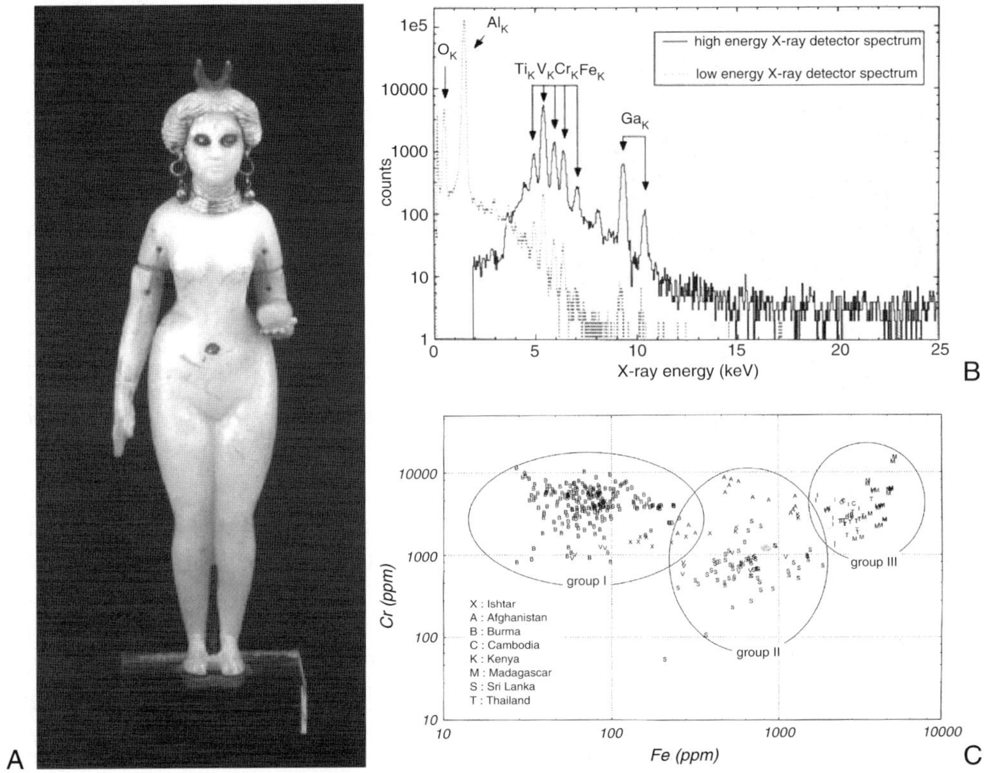


**Fig. 4.** Example of determination by PIXE of the geographical origin of a gemstone: ruby inlay on the Parthe Ishtar Statue – 300 BC (Louvre Museum) – (A) Photograph; (B) Characteristic spectra for rubies on the statue (eyes and navel); (C) Results showing that the contents of the chromium and iron are consistent with the rubies being of burmese origin (T. Calligaro and D. Bagault, C2RMF).

of the elements to be analysed and those of the matrix. So, when the target emits its characteristic spectrum in the PIXE mode, these characteristic X-rays stimulate the emission of characteristic X-rays of the lightweight elements to be analysed, without interference from the spectral lines of the heavyweight elements of the matrix.

Thus, a germanium target will produce $X_K$ photons of 9.98 keV, an energy adapted for the excitation of $X_K$ lines of copper and zinc included in a lead matrix, without interfering with the $X_L$ lines of this metal (10.45 and 10.55 keV), and thereby analysing copper and zinc.

4.3.2.5. ERDA [129]. ERDA is a technique based on the elastic diffusion of nuclei lighter than the projectile. By using an external beam of helium ions, one can determine hydrogen concentration depth profiles in gemstones like emeralds.

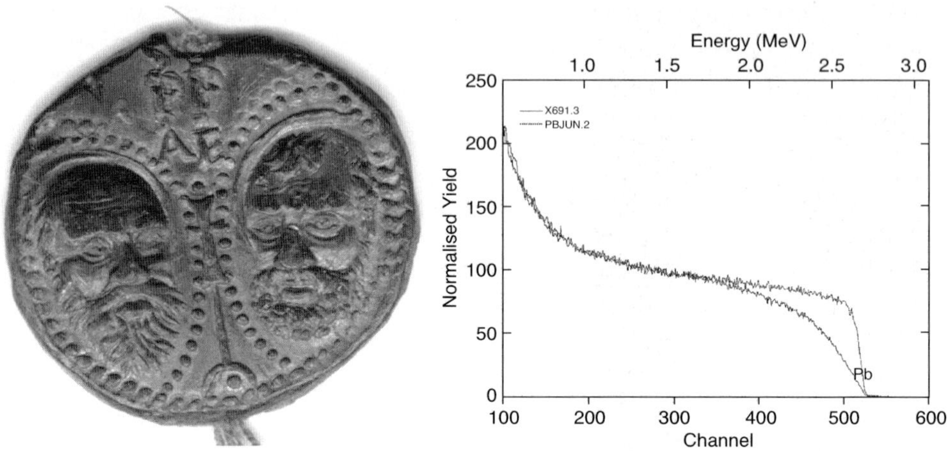


A B

**Fig. 5.** (A) Photograph of a lead Papal Bull (French National Archives, Paris); (B) The results of RBS profiles, showing different superficial alteration layer thicknesses, M. Dubus (C2RMF).

### *4.3.3. Activation analysis*

One uses nuclear reactions induced in the specimen by a particle beam (usually a neutron beam) to render certain constitutive elements of a material radioactive, permitting their analysis by identification of the radiative decay products.

The most common technique is neutron activation. The thermal neutron flux from research reactors like those operated in Berlin, Columbia, Delft, Garching, or Saclay, creates unstable nuclei in the specimen by the process of neutron capture. The resulting nuclear transitions result in $\gamma$-ray emission, K-capture, and other processes which then enables identification of the isotopic species present, and from that, of the elements such as Au, Ag, Fe, Cu, As, Co, Sn, In, rare earths, and so on. Recently the International Atomic Energy agency (IAEA) published the results of a co-ordinated research programme on the analysis of pre-Hispanic American potteries [130].

A second possibility is to use charged particle analysis produced by accelerators, like the CNRS – Orléans cyclotron [131,132], to create the nuclear reaction.

A third possibility consists in prompt gamma analysis, directly on the objects, using external collimated thermal neutron beams from research reactors, analogous to those used for neutron radiography.

### *4.3.4. Characterisation by synchrotron radiation [133–149]*

Intense, monochromatic X-ray beams emitted by synchrotron radiation sources (LURE – Orsay or ESRF – Grenoble, or Daresbury) allows structural information to be obtained from very small samples as an advanced complementary tool of X-ray classical diffraction apparatus. One example concerns the study of lead-based cosmetics used during the Pharaonic Egyptian era. At first, the study was based on laboratory X-ray diffraction,

SEM X-ray spectrometry, and FT-IR spectrometry [133–134]. With synchrotron radiation X-ray diffraction, it is possible to make measurements of crystal type on individual grains of a mixture. Samples of about 5–8 mg analysed with the synchrotron radiation permit one to show the diversity of these products: black galena (PbS), white products as laurionite (PbOHCl), phosgenite ($Pb_2CO_3Cl$), anglesite ($PbSO_4$), and cerusite ($PbCO_3$). Other significant examples concern the characterisation of pigments in lustre ceramics, the analysis of ancient bronze metal armours, or the identification of archaeological textile fibres ... [136–149].

A detailed chapter on synchrotron radiation and the techniques which may be applied to the study of artefacts of cultural heritage significance will be given in the next volume of this book series.

#### *4.3.5. X-ray diffraction [150,151]*

This is the usual method for the characterisation of crystalline structures and is used for identification of minerals, components of rocks, pigments, and alloy phases. The diffraction data are most frequently analysed using the Rietveld technique [152], which enables not only the determination of the crystal structure of components of a mixture but also the percentage composition of each phase.

Recent applications are in the distinction between hematite and goethite in the parietal paintings in the Pyrenees area, the identification of surface alteration products of silver or lead objects kept in museums, and the characterisation of micro samples (paint layers) or very small sized objects (hairs, threads).

#### *4.3.6. Neutron diffraction [153–157]*

Neutron diffraction can also be applied, using for instance a neutron spallation source facility, like the ISIS facility of the Rutherford Appleton Laboratory (UK), in order to characterise the texture of metals (cast, forged, rolled etc.), or even stones like marbles from the Villa Adriana.

#### *4.3.7. Atomic emission spectrometry*

The C2RMF Laboratory operates atomic emission spectrometry (ICP-AES) equipment, mainly dedicated to the destructive analysis of archaeological metal objects (copper alloys, lead, gold, silver). In the case of copper alloys, important methodological work has been undertaken to extend the reference data base provided by the manufacturer to include minor components (P, S, Se, Te, Ti, V, Cr, Mo, In, W, U). This now permits us to determine the place of origin of alloys [158,159]. A comprehensive study has been made on a corpus of 60 statues from the museums of Phnom-Penh (Cambodia) and Guimet (Paris) of Hindu and Buddhist Khmer art from seventh to sixteenth centuries, showing two main groups: classical bronze and lead bronze, and also determining the gold content of the statues.

#### *4.3.8. Spectro-photo-colorimetry*

A prototype of a light weight portable spectro-photo-colorimetry apparatus, using a halogen light source, quartz optical fibres for illumination and for detection (100 fibres 50 μm diameter) coupled to an achromatic quartz doublet lens (diameter of the spot on the object 6 mm) and a spectrometer, dedicated to laboratory and *in-situ* measurement (museum collections, restoration workshops, frescoes, parietal paintings, polychromatic statues, etc.) has been developed [160–166].

The spectrometer comprises a diffraction pattern and a CCD array of 1100 pixels, giving a spectrum from 350 to 850 nm. Illumination and light detection are made at the same angle (nominal 22° from the normal incidence).

Different modes are used in order to examine different optical and surface properties of coloured layers, such as:

- the spectrum of reflected visible light, *i.e.* classical colour characterisation of a coloured layer, from which classical CIELAB data can be produced (hue, brightness, and chroma) (this has application in the recognition of papers, pigments, pastels);
- the modification of these parameters for coloured layers submitted to weathering tests (light, UV, moisture, corrosive atmosphere, dust, combination of these factors);
- the roughness index, through the measurement of the widening of the distribution of the reflected light flux, as the illumination angle varies;
- the characterisation of glazed layers. Study of the correlation between optical models of multiple transparent coloured layers light response and experimental spectra.

### *4.3.9. Infrared spectrometry [167–170]*

Infrared spectrometry and derived techniques (FT-IR) are widely used for cultural heritage materials analysis. An Infrared & Raman Users Group (IRUG) has been created (1994). Much information relative to these techniques, including database and online bibliography, can be found on the website: http://www.irug.org

### *4.3.10. Raman spectrometry*

The C2RMF has recently acquired a Raman spectrometer, the main characteristics of which are:

- 2 laser sources: red (He–Ne, $\lambda = 632$ nm) and green (diode, $\lambda = 532$ nm);
- adjustable power between 1 and some 100 $\mu$W;
- possibility of examination with an internal chamber ($200 \times 300 \times 50$ mm) and with an external beam;
- spot diameter on the object 1–1.5 $\mu$m;
- viewing camera.

The first applications concern pigments, gems, and lead glasses characterisation [171,172].

One must mention very interesting results concerning the characterisation and the determination of alteration index of natural fibres of ancient textiles (ensigns, sails, banners, and the like) using simultaneously IR and Raman spectrometry [167–170].

This technique is probably one of the more promising methods of cultural heritage materials' identification. Many research teams and manufacturers are presently publishing protocols and results covering a broad spectrum of applications [173–183].

For general information relative to this technique, one can also consult the IRUG website.

*4.3.11. Laser-induced spectrometric techniques*

There are several laser-induced spectroscopic techniques, which can be used for the diagnosis of cultural heritage artefacts. Some are strictly non-destructive ones, do not require sample preparation, and can be used *in situ*, and some introduce small ablation craters in the examined objects.

Laser-Induced Fluorescence (LIF), Laser-Induced Breakdown Spectroscopy (LIBS), and Laser Ablation – Inductively Coupled Plasma – Mass Spectrometry (LA-ICP-MS) have been extensively used, not only for the analysis of pigments and binding media of artworks, but also for determining the degree of ageing and oxidation or polymerisation processes. Nowadays, portable workstations are available for incorporating these techniques, which provide capability for *in situ* analysis, without the need of separate sampling. In particular, LIBS presents several interesting possibilities for elemental and in-depth analysis. As it has been demonstrated, LIBS may be combined with cleaning applications, using lasers or other conventional means, for monitoring and controlling the cleaning process [184–188].

*4.3.12. Nuclear magnetic resonance (NMR) imaging*

NMR is well known for being routinely used in medical diagnosis to make soft tissue density scans. The technique is also commonly used for the industrial measurement of fat/aqueous compounds ratio in the agro-food industry.

Recently, new developments permit the making of one-side access depth profile of thin 2D objects, and discriminate between different thin layers of organic compounds [189–191].

*4.3.13. Gas chromatography*

Gas chromatography, either coupled or not coupled to mass spectrometry (GC-MS) is used for the analysis of organic materials such as paint layers, oils, media, varnishes, lacquers, archaeological residues like glues, adhesives, tars, resins, honeys, waxes, foods, beverages, and their degradation products [192–199].

A recent application was the analysis of an adhesive on a Hallstatt period (Seventh century BC) iron spade from a grave of north east France. Birch tar residue was identified from the adhesive.

A Users' Group for Mass Spectrometry and Chromatography (MaSC) has been recently created (2003). Much information relative to these techniques can be found on the website http://www.mascgroup.org

*4.3.14. Miscellaneous*

There are two very interesting initiatives which take advantage of many of the above mentioned techniques. The first one is managed by M. Derrick (Museum of Fine Arts – Boston) and consists of a "Conservation & Art Material Encyclopaedia Online" (CAMEO) website, which gives a significant amount of basic information on materials and techniques (*cf.* Appendix 2 for the website). The second one is managed by N. Eastaugh *et al.* and is a "Pigmentum Project" which leads into books [200,201] and has a website (*cf.* Appendix 2).

## 4.4. Dating

### *4.4.1. Thermoluminescence dating [202]*

The same principle of thermoluminescence of crystalline structures exposed to ionising radiation that is used for health physics dosimetry is also used for artefact dating purpose.

Using this technique, one can estimate the time elapsed since the last firing of a crystalline structure: quartz, feldspar, zircon, etc., included in pottery, terracotta, architectural elements, ceramics, oven elements, foundry cores, burned stones from fireplaces, flint tools or scrape parts, etc. This permits age estimation of the objects, or may eventually lead to the detection of a forgery [203].

The age range for pottery and other ceramics covers the entire period in which these materials have been produced. The typical range for burnt flint, stone or sediment (burnt or not) is from about 100 to 300 000 years. The error limits on the dates obtained are typically in the range ±3 to ±8%, although recent technical developments now allow luminescence measurements to be made with a precision of ±1 to ±2% in favourable circumstances.

One can usefully consult the website http:www.aber.ac.uk/ancient-tl for details.

### *4.4.2. Carbon-14 dating*

Substances of living origin are dated by the measurement of the isotopic composition in $^{14}C$ (period (half-life) of 5730 years) of the constitutive carbon. The limit of the age range is approximately 45 000 years.
Two techniques are customarily employed:

1. Counting of the β radioactivity [204]. A recent application is the dating of the charcoal used for the sketches of rhinoceros in the Chauvet cave (South of France): 30 800 to 32 400 years ± 650 years, measured by the Centre de Datation par le Radiocarbone (Université de Lyon). Information about this method can be obtained on the website http://www.c14dating.com
2. By particle acceleration, followed by mass spectrometry (AMS). See [205] for a description of AMS and its uses. A new French national equipment dedicated to this technique was inaugurated in April 2004 in the CEA/Saclay Research Centre.

### *4.4.3. Dendrochronology*

This technique permits one to establish a chronology, through the identification of the sequence of tree-ring widths, for a particular tree species and a geographical area, which is the signature of the succession of seasons and years. It should be noted that this technique can be extended to cover all organic processes where growth takes place on an annual basis. The dating of ancient coral is a recent extension of the technique.

Other dating techniques such as electron paramagnetic spin resonance (ESR) and lead isotopic composition are also used.

# 5. CONCLUSION

This chapter does not pretend to be exhaustive with respect to the techniques of analysis chosen. The purpose is just to attract attention on the diversity of techniques and the various issues to be solved, so that we may better understand our common cultural heritage, and to build rational basis for its conservation for the future generations.

## ACKNOWLEDGEMENTS

My sincere thanks go to all my former colleagues of CEA/Saclay (R. Hours, G. Courtois, A. Lemonnier, B. Rattoni, G. Bayon) and C2RMF-Paris (J.P. Mohen, M. Menu, F. Dijoud, M. Aucouturier, D. Bagault, T. Borel, A. Bouquillon, J. Castaing, D. Bourgarit, T. Calligaro, J.C. Dran, A. Duval, M. Dubus, M. Elias, A. Fortune, O. Guillon, M.O. Kleitz, B. Mille, C. Moulherat, E. Ravaud, M. Regert, J. Salomon, D. Vigears) and also to all the LabS TECH and EU-ARTECH team members.

## APPENDIX 1: SOME NATIONAL CULTURAL HERITAGE INSTITUTIONS

Belgium – Institut Royal du Patrimoine Artistique (IRPA) – Bruxelles, http://www.kikirpa.be/www2/

France

- Centre de Recherche et de Restauration des Musées de France (C2RMF) – Paris & Versailles, http://www.c2rmf.fr/
- Laboratoire de Recherche des Monuments Historiques (LRMH) – Champs sur Marne, http://www.lrmh.fr/
- Centre de Recherche sur la Conservation des Documents Graphiques (CRCDG) – Paris, http://www.crcdg.culture.fr/

Germany

- Rathgen Forschungslabor – Berlin, http://www.smb.spk-berlin.de/fw/rf/
- Bayerisches Landesamt für Denkmalpflege – Munich, http://www.blfd.bayern.de/blfd/

Italy

- Istituto Centrale del Restauro – Rome, http://www.icr.arti.beniculturali.it/
- Opificio delle Pietre Dure – Florence, http://www.opificio.arti.beniculturali.it/ita/home.htm
- Istituto Centrale di Patologia del Libro – Rome, http://www.patologialibro.beniculturali.it/

Netherlands

- Instituut van Collectie Nederland (ICN) – Amsterdam, http://www.icn.nl/Dir003/ICN/CMT/Homepage.nsf/index2.html?readform

Spain

- Instituto del Patrimonio Historico Español – Madrid, http://www.cultura.mecd.es/patrimonio/iphe/institutoPatrimonioHistorico

United Kingdom

- Scientific Departments of the British Museum – London, http://www.thebritishmuseum.ac.uk/science/
- The National Gallery London, http://www.nationalgallery.org.uk/
- The Victoria and Albert Museum – London, http://www.vam.ac.uk/res_cons/index.html
- The Tate Gallery – London, http://www.tate.org.uk/home/default.htm

Canada

- Canadian Conservation Institute – Ottawa, http://www.cci-icc.gc.ca/

United States

- Smithsonian Center for Materials Research and Education – Smithsonian Institution – Washington – DC, http://www.si.edu/scmre/
- Getty Conservation Institute – Los Angeles – CA, http://www.getty.edu/conservation/

Australia

- Australian Institute for the Conservation of Cultural Material (AICCM) – Canberra http://www.aiccm.org.au/aiccm/home/

Japan

- National Research Institute for Cultural Properties (TOBUNKEN) – Tokyo, http://www.tobunken.go.jp/

## APPENDIX 2: WEBSITES OF INTEREST IN THE DOMAIN "SCIENCE AND TECHNOLOGY" AND "CULTURAL HERITAGE"

AAT – Art and Architecture Thesaurus Online (Getty), http://www.getty.edu/research/conducting_research/vocabularies/aat/index.html

ATAM – Ancient Technologies and Archaeological Materials (U. Illinois Urbana Champaign), http://www2.uiuc.edu/unit/ATAM/

Ausbildungstätten für Restauratoren, http://home.rol3.com/~u0369118/hochsch.htm

Signets de la Bibliothèque Nationale de France, http://signets.bnf.fr/

CAMEO – Conservation & Art Material Encyclopaedia Online (MFA Boston), http://signets.bnf.fr/

CHIN – Canadian Heritage Information Network, http://www.chin.gc.ca/

CoOL – Conservation OnLine (Stanford U.), http://palimpsest.stanford.edu/

Courses & Education in Heritage Conservation (Robert Gordon U. Aberdeen), http://www2.rgu.ac.uk/schools/mcrg/stuni.htm

Cultural Heritage Search Engine, http://www.culturalheritage.net/

e Preservation Science (U. Ljubljana), http://rcul.uni-lj.si/~eps/index.html

EachMed – Agenzia Europea e Mediterranea per i Beni Culturali, http://213.92.94.10/portale_pfbc/home.asp

ECPA – European Commission on Preservation and Access, http://www.knaw.nl/ecpa/

European Cultural Heritage Network (Fachhochschule Köln), http://www.echn.net/echn/

EMII – European Museums' Information Institute, http://www.emii.org/

IICROM – International Centre for the Study of the Preservation and Restoration, of Cultural Property, http://www.iccrom.org/eng/news/iccrom.htm

ICOM-CC – International Council of Museums – Committee for Conservation, http://www.icom-cc.org/
IIC – International Institute for Conservation of Historic and Artistic Works, http://www.iiconservation.org/
ILAM – Instituto Latinoamericano de Museos, http://www.ilam.org/
INCCA – International Network for the Conservation of Contemporary Art, http://www.incca.org/
IAQ – Indoor Air Quality in Museums and Archives, http://www.iaq.dk/
Kunst als Wissenschaft – Wissenschaft als Kunst, http://www.kunst-als-wissenschaft.de/de/index.html
New York Conservation Foundation, http://www.nycf.org/
OCIM – Office de Coopération et d'Information Muséographiques (U. Bourgogne Dijon), http://www.ocim.fr/sommaire/
Pigmentum Project, http://www.pigmentum.org/
Red Tematica de Patrimonio Historico y Cultural (CSIC – Spain), http://www.rtphc.csic.es/
University of Delaware Internet Resources for Art Conservation, http://www2.lib.udel.edu/subj/artc/internet.htm.
WAAC – Western Association for Art Conservation (U. Stanford), http://palimpsest.stanford.edu/waac/

## APPENDIX 3: SOME PUBLICATIONS OF INTEREST IN THE DOMAIN "SCIENCE AND TECHNOLOGY" AND "CULTURAL HERITAGE"

Archaeometry, http://www.rlaha.ox.ac.uk/archy/archindx.html
Journal of archaeological science, http://www.sciencedirect.com/science/journal/03054403
Journal of Conservation and Museum Studies, http://www.jcms.ucl.ac.uk/
Journal of cultural heritage, http://www.sciencedirect.com/science/journal/12962074
Kermes, http://www.nardinirestauro.it/index_base.asp?zoom=homepage2&idCanale=2
Restauro, http://www.restauro.de/frames.htm
Studies in conservation, http://www.jxj.com/sinc/index.php
Techné, http://www.c2rmf.fr/pages/page_id18439_u1l2.htm

## APPENDIX 4: QUESTIONS TO BE SOLVED BY RADIOGRAPHY, SOME EXAMPLES

For certain categories of objects and/or materials, one has to solve different problems and solutions. Let us illustrate this through some generic examples.

### A. Paper, support of drawing or text

Visualisation of the texture of the paper, of local variation of mass per unit area, of the watermarks, and whatever drawings or texts are on it.

*Solution:* Use beta radiography or radiography with secondary electrons.

### B. Easel paintings

The classical structure of an easel painting is a multilayer one:

- plane support: canvas (linen, hemp, cotton), wood (panel of different species, cut according to different manners), metal (copper), stone (obsidian);
- preparation layer (chalk, gypsum (or gesso), lead white);
- animal glue;
- eventual underdrawing (carbon black);
- paint layer: adjacent and/or superimposed spots of organic or mineral pigments suspended in organic media;
- varnish.

One wants to determine the texture of the backing and of the paint layer, the characteristic "touche" of the author, the eventual existence of an underlying painting, the pentimenti, the alterations, the restorations, the modifications of frame, the linings, and the transpositions.

*Solution:* Use low energy X-ray radiography and/or electron emission radiography, and possibly laminography.

### C. Enamels

Glassy layer loaded with mineral pigments on a metallic backing (generally copper alloy). One has to determine contrasts in the composition of the enamel layer and its alterations.

*Solution:* Use X-ray radiography, sometimes at different energies (selective filtration), and electron emission radiography.

### D. Wood

The goal is to visualise the characteristic texture of the wood species, the assembly techniques, the alterations (flaws, brittleness areas), the infestations, the restorations, and the eventual metallic inserts.

*Solution:* Use X-ray radiography.

### E. Stone

One wants to determine the texture of the stone material, its homogeneity, the eventual cleavage planes, the restorations, and the eventual metallic inserts.

*Solution:* Use high energy X-ray radiography or gamma radiography.

### F. Foundry (metal)

The goal is to visualise the manufacturing process, the assembly techniques, the defects and alterations (porosity, bubbles, flaws, corroded areas …), the restorations, and the eventual metallic inserts.

*Solution:* Use high energy X-ray radiography, radioscopy, or gamma radiography.

## REFERENCES


[1] C. Brandi, *Teoria del restauro*, Einaudi, Torino, 2000.
[2] C. Brandi, *Il restauro – Teoria e pratica*, Editori Riuniti, Roma, 1996.
[3] M. Hours, *La vie mystérieuse des chefs d'oeuvre – La science au service de l'art*, RMN, Paris, 1980.
[4] P. Nora, *Science et conscience du patrimoine – Actes des entretiens du patrimoine – Paris 1994,* Fayard, Paris, 1997.
[5] J.P. Mohen, *L'Art et la Science*, Gallimard, Paris, 1996.
[6] J.P. Mohen, *Les Sciences du Patrimoine*, Odile Jacob, Paris, 1999.
[7] Anonymous (UNESCO), *La conservation préventive*, Museum International, **51** (1999) 1.
[8] J. Ashley-Smith, *Risk Assessment for Object Conservation*, Butterworth-Heinemann, London, 1999.
[9] M. Bacci, C. Cucci, A.A. Mencaglia, A.G. Migniani, S. Porcinai, Calibration and use of photosensistive materials for light monitoring in museums: blue wool standard 1 as a case study, *Studies in Conservation*, **49** (2004) 2, 85–94.
[10] J.M. Banks, *Guidelines for preventive conservation*, National Library of Canada, Ottawa, 1987.
[11] M. Berducou, La restauration, quel choix? Dérestaurer, restaurer, restituer, *Techné*, **13** (2000) 14.
[12] S. Corr, *Caring for Collections*, The Heritage Council, Kilkenny, 2000.
[13] G. de Guichen, La conservation preventive: simple mode ou changement profond, *Museum International*, **51** (1999) 1.
[14] D. Guillemard, C. Laroque, *Manuel de conservation preventive: gestion et contrôle des collections*, OCIM, Dijon, 1999.
[15] P. Hatchfield, *Pollutants in the Museum Environment: Practical Strategies for Problem Solving in Design, Exhibition and Storage*, Archetype Publishing, London, (2002).
[16] C.V. Horie, *Materials for Conservation*, Butterworth-Heinemann, London, 1987.
[17] B. Lavédrine, J.P. Gandolfo, S. Monod, *Les collections photographiques. Guide de conservation préventive*, ARSAG, Paris, 2000.
[18] A. Levillain, P. Markarian, C. Rat, P. Mairot *et al.*, *La conservation preventive des collections. Fiches pratiques à l'usage des personnels de musée*, OCIM, Dijon, 2002.
[19] T. Padfield, Conservation Physics, http://www.padfield.org/tim/cfys
[20] T. Padfield, P.K. Larsen, Designing museums with a naturally stable climate, *Studies in Conservation*, **49** (2004) 131–137.
[21] C. Pearson, *Conservation of Marine Archaeology Objects*, Butterworth-Heinemann, London, 1988.
[22] E. Pye, *Caring for the Past: Issues in Conservation for Archaeology and Museums*, James and James, London, 2000.
[23] J. Tétrault, *The Good, the Bad and the Ugly*, http://www.iaq.dk/papers/good-bad-ugly.htm
[24] G. Thompson, *The Museum Environment*, Butterworth-Heinemann, London, 1993.
[25] S. Wuelfert, *Internet Teaching in Conservation*, http://surf.agri.ch/wuelfert
[26] H. Yngvason, *Conservation and Maintenance of Contemporary Public Art*, Archetype Publishing, London, 2002.
[27] G. Demortier, A. Adriaens, *Ion beam study of art and archaeological objects. A contribution by members of the COST G1 action*, Report EUR 19218, 2000.
[28] J.H. Townsend, K. Eremin, A. Adriaens, Conservation science 2002, in: *Proceedings of a COST G8 meeting, Edinburgh (May 2002)*, Archetype Publishing, London, 2003.
[29] B. Brunetti, A.M. Johansson, *Science and technology for the conservation of the European cultural heritage – Research infrastructures*, Report EUR 20483, 2003.
[30] *Proc. Workshop on Binding Media Identification in Art Objects*, ICN Amsterdam, March 24–28, 2003.
[31] *Proc. Workshop on the Painting Technique of Pietro Vannucci called "il Perugino", Perugia*, Nardini, Florence April 14–15, 2003.
[32] *Proc. Workshop on silicon based products in the sphere of cultural heritage*, ICCROM, Rome, April 29–30, 2004.
[33] *Proc. Workshop on novel technologies for digital preservation information processing and access to cultural heritage collections*, Ormylia, May 21–22, 2004.
[34] *Proc. EU-ARTECH workshop on Raphael's painting technique: working practices before Rome, The National Gallery, London*, Archetype Publication, London, November 11, 2004 (to be published).

[35] E. Ciliberto, G. Spoto, *Modern analytical methods in art and archaeology*, Wiley Interscience, New York, 2000.
[36] D.C. Creagh, D.A. Bradley, *Radiation in art and archaeometry*, Elsevier, Amsterdam, 2000.
[37] K. Janssens, R. Van Grieken, *Non-destructive Micro Analysis of Cultural Heritage Materials*, Elsevier, Amsterdam, (2004).
[38] E.S. Pardo, *Tecnicas de diagnostico applicadas a la conservacion de los materials de construccion en los edifices historicos*, Junta de Andalucia, Sevilla, 1996.
[39] J.H.H. de Graaf, W.G.H. Roelofs, M. van Bommel, *The Colourful Past: the Origin, Chemistry and Identification of Natural Dyestuffs*, Archetype Publishing, London, 2004.
[40] N. Odegaard, S. Carroll, *Material Characterisation Tests for Objects of Art and Archaeology*, Archetype Publishing, London, 2000.
[41] *Proc. Conference on X-ray spectrometry – EXRS 2004*, Alghero, June 6–11, 2004 (to be published).
[42] F. Mairinger, The infrared examination of paintings, in: D.C. Creagh and D.A, Bradley (Eds.), *Radiation in Art and Archaeometry*, Elsevier, Amsterdam, 2000, pp. 40–76.
[43] D. Vigears, private communication, 2004.
[44] C. Moulherat, *Archéologie des textiles protohistoriques – Exemple de la Gaule celtique*, Thesis, University of Paris 1, 2002.
[45] C. Moulherat, M. Tengberg, J.F. Haquet, B. Mille, First evidence of cotton at Neolithic Mehrgarh, Pakistan. Analysis of mineralised fibres from a copper bead, *J. Archaeol. Sci.*, **29** (2002) 12, 1393–1401.
[46] M. Hours, La radiographie des oeuvres d'art, in Idées et Découvertes Encyclopaedia Universalis, Paris, 1978, pp. 240–243.
[47] M. Hours, *Les secrets des chefs d'oeuvre*, R. Laffont, Paris, 1964.
[48] A. Gilardoni, L. Ascani Orsini, R. Ascani Orsini, M. Taccani Gilardoni, S. Taccani, X-rays in art – I raggi X nell'arte, (Second ed.), Gilardoni, Mandello Lario, 1984.
[49] D. Graham, T. Eddie, *X-ray Techniques in Art Galleries and Museums*, A. Hilger, Bristol, 1985.
[50] J. Lang, *Radiography of Cultural Materials*, Butterworth-Heinemann, London, 1997.
[51] J.L. Boutaine, T. Borel, E. Ravaud, La radiographie au service de l'étude des oeuvres du patrimoine culturel, in: *Proc. 3ème séminaire franco-allemand sur le contrôle non-destructif – COFREND – DGZfP – Forbach*, September 9–10, 1999.
[52] P.A. Ruault, *Radiologie Industrielle*, Vol. 2, Publications de la Soudure Autogène, Paris, 1991.
[53] R. Halmshaw, *Industrial Radiology: Theory and Practice*, second ed., Chapman and Hall, London 1995.
[54] D. Faber, *Eine neue Anwendung der Röntgenstrahlen*, Die Umschau, **12**, 1914.
[55] A. Burroughs, *Notes on the principles and process of X-ray examination of paintings*, Smithsonian report for 1927, Washington DC, 1928.
[56] K. Wehlte, *Röntgenstrahlen im Dienst der Mahltechnik*, Kunst und Wirtschaft, 13, 1931.
[57] J. Wilde, *L'examen des tableaux à l'Institut Holzknecht de Vienne*, Mouseïon, 16, 18, 1931.
[58] C. Wolters, Die Bedeutung der Gemälde durch Leuchtung mit Röntgenstrahlen für die Kunstgeshichte, Prestel Verlag, Frankfurt am Main, 1938.
[59] F. Mairinger, Untersuchungen von Kunstwerken mit sichtbaren und unsichtbaren Strahlen, Bildhefte der Akademie der Bildenden Künste in Wien, Wien, **8** (1977) 9.
[60] L. Mucchi, A. Bertuzzi, Nella profondita' dei dipinti – La radiografia nell'indagine pittorica, Electa, Milano, 1983.
[61] E. Martin, E. Ravaud, La radiophotographie des peintures de chevalet, *Techné*, **2** (1995), 158–164.
[62] J.R.J. Van Asperen de Boer, An introduction to the scientific examination of paintings, in: *Scientific examination of early Netherlandish paintings – Nederlands Kunsthistorisch Jaarboek*, Fibula, van Dishoeck Bussum, 1976.
[63] S. Rees-Jones, Paintings and optical absorption, X-ray absorption, *Bulletin de l'IRPA*, **15** (1975), 326–336.
[64] E. Ravaud, Evaluation de la méthode radiographique dans l'identification des essences de bois de panneaux peints, in: *Proc. ICOM-CC 12th Triennal Meeting*, Lyon, September 1999.
[65] H. Van Hugten, Weichstrahlradiographie zum Beispiel bei Papier, in: *Proc. Symposium Zerstörungsfreie Prüfung von Kunstwerken, Berlin*, DGZfP Berlin, November 19–20, 1987.
[66] P. Baldelli, M. Gambaccini, M. Milazzo, F. Petrucci, M. Scotti, K-edge differential radiography of art paintings, in: *Proc. EXRS 2004 European Conf. on X-ray Spectrometry,* Alghero, June 2004.
[67] F. Drilhon, L'examen radiographique de sculptures en cire du XVI au XIX$^{e}$ siècles, in: *Proc. ICOM-CC 7th Triennal Meeting*, Copenhague, September 1984.

[68] C. Carr, Advances in ceramic radiography and analysis, applications and potentials, *Jl. Archael. Sc.*, **17** (1990) 1, 13–34.
[69] T. Borel, La radiographie des objets d'art, *Techné*, **2** (1995), 146–151.
[70] H. Althofer, *La radiologia per il restauro delle opere d'arte moderne e contemporanee*, Nardini, Fiesole, 1998.
[71] J.L. Boutaine, J. Irigoin, A. Lemonnier, La radiophotographie dans l'étude des manuscrits, in: *Proc. Colloque International du CNRS N° 458 – Les Techniques de Laboratoire dans l'Etude des Manuscrits*, CNRS, Paris, September 1972, pp. 159–176.
[72] A. de La Chapelle, A. Le Prat, Les relevés de filigranes. in: *La Documentation Française*, Paris, 1996.
[73] C.F. Bridgman, S. Keck, H.F. Sherwood, The radiography of panel paintings by electron emission, *Studies in Conservation*, **3** (1958) 4, 175–182.
[74] D. Schnitger, E. Ziesche, Elektronenradiographie von Wasserzeichen, in: *Proc. Symposium Zerstörungsfreie Prüfung von Kunstwerken*, Berlin, DGZfP Berlin, November 19–20, 1987.
[75] B. Quette, I. Biron, T. Borel, Deux émaux cloisonnés chinois du musée des arts décoratifs, *La Revue du Louvre*, **2** (1998), 42–46.
[76] B. Bourgeois, J.L. Boutaine, B. Rattoni, Radiographie et restauration de marbres antiques: l'exemple du Musée du Louvre, in: *Proc. ICOM-CC 11th Triennal Meeting*, Edimburgh, September 1996, pp. 793–797.
[77] T. Borel, Applications de la radiographie X et γ à l'étude des objets métalliques d'art et d'archéologie – Emissiographie, in: *A la recherche du métal perdu* – Catalogue de l'exposition de Guiry-en-Vexin, 1999, pp. 54–60
[78] O. Sicardy, *Application de quelques techniques de contrôle non destructif à la caractérisation des roches calcaires*, Thèse, Ecole Centrale Paris, July 1985.
[79] H. Berger, *Neutron radiography. Methods, capabilities and applications*, Elsevier, Amsterdam, 1965.
[80] G. Bayon, Les contrôles non destructifs et la neutronographie, in: *Proc. 7ème Conf. Europ. sur les CND Nice*, 1994, p. 1043.
[81] Anonymous, *Art and autoradiography: Insight into the genesis of paintings by Rembrandt, Van Dyck and Vermeer*, The Metropolitan Museum of Art, New York, 1981.
[82] C.O. Fischer et al., Autoradiography of paintings after neutron activation at a cold neutron guide, *Kerntechnik*, **51** (1987) 1, 9–13.
[83] C. Barry, La Tour and autoradiography, in: Philip Conisbee (Ed.) *Georges de la Tour and his World.* National Gallery of Art Washington DC, Yale University Press, New Haven CT, 1996.
[84] C.O. Fischer et al., Autoradiography of paintings, in: *Proc. 5th Conf. on Neutron Radiography*, Berlin, DGZfP, 1996.
[85] P. Targowski, B. Rouba, M. Wojtkowski, A. Kowalczyk, The application of optical coherence tomography to non-destructive examination of museum objects, *Studies in Conservation*, **49** (2004) 2, pp. 107–114.
[86] V. Tornari, A. Bonarou, V. Zafiropulos, C. Fotakis and M. Doulgeridis, Holographic applications in evaluation of defect and cleaning procedures, *Journal of Cultural Heritage*, **1** (2000), S325–S329.
[87] N.P. Avdelidis, A. Moropoulou, Applications of infrared thermography for the investigation of historic structures, *Journal of Cultural Heritage*, **5** (2004) 1, 119–127.
[88] M. Menu, G. Padeletti, M. Stuke, P. Vandiver, Materials aspects of art characterization, conservation and restoration, ERMS 2004 Conference, Strasbourg, *Applied Phy. A*, **79** (2004), 161–400.
[89] E.S. Lindgren, Special millennium issue on cultural heritage, *X-ray Spectrometry*, **29** (2000) 1.
[90] W. Devos, L. Moens, A. Von Bohlen, R. Klockenkämper, Ultra-microanalysis of inorganic pigments on painted objects by total-reflection X-ray fluorescence analysis, *Studies in Conservation*, **40** (1995), 153–162.
[91] L. Moens, W. Devos, R. Klockenkämper, A. von Bohlen; Application of TXRF for the ultra micro analysis of artists' pigments, *J. Trace and Microprobe Techniques*, **13** (1995) 2, 119–139.
[92] R. Klockenkämper, A. Von Bohlen, L. Moens, Analysis of pigments and inks on oil paintings and historical manuscripts using total reflection X-ray fluorescence spectrometry, *X-Ray spectrometry*, **29** (2000) 1, 119–129.
[93] *Proc. ECAART-8 2004, 8th European Conference on Accelerators in Applied Research and Technology*, Paris, September 20–24, 2004, (to be published in NIM).
[94] T. Calligaro, J.C. Dran, M. Dubus, B. Moignard, L. Pichon, J. Salomon, P. Walter, Les objets de musée sous le projecteur d'AGLAE, *Techné*, 13–14 (2001), 41–48.
[95] T. Calligaro, J.C. Dran, J. Salomon, Quand un accélérateur de particules sonde les œuvres d'art, *Découverte*, **298** (2002), 8–17.

[96] J.C. Dran, T. Calligaro, J. Salomon, Particle-induced X-ray emission, in: *Modern Analytical Methods in Art and Archaeology*, Wiley Interscience, New York, 2000, pp. 135–166.

[97] T. Calligaro, J.C. Dran, E. Ioannidou, B. Moignard, L. Pichon, J. Salomon, Development of an external beam nuclear microprobe on the AGLAE facility of the Louvre museum, *Proc. IBA-14*, Dresden, July 1999, in: NIM-B, **161–163** (2000), pp. 328–333.

[98] T. Calligaro, J.C. Dran, B. Moignard, L. Pichon, J. Salomon, P. Walter, Ion beam analysis with external beams: recent set-up improvements, *Proc. ECAART 7*, Guildford, in: NIM-B, **188** (2002), pp.135–140.

[99] T. Calligaro, J.P. Poirot, G. Querré, Trace element fingerprinting of jewellery rubies by external beam PIXE, in: NIM-B, **150** (1999) pp. 628–634.

[100] T. Calligaro, J.C. Dran, J.P. Poirot, J. Salomon, PIXE/PIGE characterisation of emeralds using an external micro-beam, *Proc. IBA-14*, Dresden, July 1999, in: NIM-B, **161–163** (2000), pp. 769–774.

[101] T. Calligaro, J.C. Dran, J.P. Poirot, G. Querré, Ion Beam Analysis techniques: a powerful set of tools for the identification and sourcing of ancient gems, in: *Proc. III Congreso Nacional de Arqueometria*, Sevilla, September 1999, pp. 197–206.

[102] A.M.B. Olsson, T. Calligaro, S. Colinart, J.C. Dran, N.E.G. Lovestam, B. Moignard, J. Salomon, Micro-PIXE analysis of an ancient Egyptian papyrus: Identification of pigments used for the "Book of the Dead", NIM-B, **181** (2001), pp. 707–714.

[103] M. Eveno, M. Dubus, A. Duval, M. Elias, Study of a 9th century illuminated manuscript from St Amand by PIXE, spectrophotometry and X-ray diffraction, in: *Non-Destructive Testing and Micro-Analysis for the Diagnostic and Conservation of the Cultural and Environmental Heritage, Proc. Art 2002*, Antwerpen, 2002.

[104] C. Remazeilles, T. Calligaro, J.C. Dran, L. Pichon, V. Quillet, J. Salomon, PIXE elemental mapping on original manuscripts with an external micro-beam. Application to manuscripts damaged by iron-gall ink corrosion, NIM-B, **181** (2001), 681–687.

[105] A. Duval, PIXE analysis of metal point drawings. in: *Proc. 6th International Conference on "Non-Destructive Testing and Microanalysis for the Diagnostics and Conservation of the Cultural and Environmental Heritage",* vol. 2, Roma, May 17–20, 1999, 1007–1022.

[106] I. Reiche, A. Berger, A. Duval, W. Görner, H. Guicharnaud, H. Merchel, M. Radke, J. Riederer, H. Riesemeier, Non-destructive investigation of Dürer's silver point drawings by PIXE and SR-XRF, in: *Non-Destructive Testing and Micro-Analysis for the Diagnostic and Conservation of the Cultural and Environmental Heritage, Proc. Art 2002*, Antwerpen, June 2002.

[107] L. Montalbano, G. Casu, A. Duval, C. Frosinini, H. Guicharnaud, Italian metal point drawings: international studies of the artistic technique, in: *Proc. 13th ICOM-CC Triennial Meeting*, Rio de Janeiro, James and James, London, 2002.

[108] M.F. Guerra, T. Calligaro, J.C. Dran, C. Moulherat, J. Salomon, Development of a PIXE, PIGE and PIXE-XRF Combination for the Analysis of the Gold from the First Empire of the Steppes, *Geoarcheological and Bioarchaeological Studies* (to be published).

[109] J. Opitz-Coutureau, A. Denker, C. Couzon, R. Denk, M. Griesser, H. Winter, E. Nagel, Zerstörungsfreie Elementanalyse durch Hochenergie-PIXE an mittelalterlichen „Wiener Pfennigen" und mongolischen Bronze- Fundmünzen aus Karakorum, in: *Proc. Symposium „Numismatics and Technology: Questions and Answers"*, Kunsthistorisches Museum, Vienna, April 25–26, 2003 (to be published).

[110] A. Zucchiatti, A. Bouquillon, J.R. Gaborit, B. Moignard, J. Salomon, Study of Italian renaissance sculptures using an external beam nuclear microprobe, *Proc. IBA-14*, Dresden, July 1999, in: NIM-B, **161–163** (2000), pp. 699–703.

[111] A. Zucchiatti, A. Bouquillon, G. Lanterna, F. Lucarelli, P.A. Mando, P. Prati, J. Salomon, M.G. Vaccari, PIXE and μ-PIXE analysis of glazes from terracotta sculptures of the della Robbia workshop, *Proc. ECAART 7*, Guildford, in: NIM-B, **188** (2002), pp. 358–363.

[112] A. Bouquillon, J. Castaing, J.R. Gaborit, E. Vartanian, A. Zink, A. Zucchiatti, Etude des oeuvres robbiesques au centre de Recherche et de Restauration des musées de France, in: *Les Della Robbia, sculptures en terre cuite émaillée de la Renaissance italienne*, RMN, Paris, 2002.

[113] E. Darque-Ceretti, D. Hélary, M. Aucouturier, An investigation of gold/ceramic interfaces, *Gold Bulletin,* **35** (2002) 4, 118–129.

[114] E. Darque-Ceretti, M. Aucouturier, Secondary ion mass spectrometry. Application to archaeology and art objects, in: K. Janssens and R. van Grieken (Eds.), *Non-destructive Microanalysis of Cultural Heritage Materials*, Elsevier, Amsterdam, 2004.

[115] D. Hélary, Etude de couches dorées sur matières vitreuses. Application aux tesselles à feuille d'or et aux céramiques glaçurées à décors de lustres dorés, Thesis, Ecole Nationale Supérieure des Mines de Paris, 2003.
[116] D. Chaulet, A. Bouquillon, P. Le Coustumer, J.H. Thomassin, Mobility or immobility of lead during glass surfaces alteration in different environments, in: *Proc. GS-2000*, Venezia, November 2000, pp. 131–133.
[117] C. Bonnet, A. Bouquillon, S. Turrell, V. Deram, B. Mille, J. Salomon, J.H. Thomassin, C. Fiaud, Alteration of lead silicate glasses due to leaching in heated acid solutions, *J. Non crystalline solids* (to be published).
[118] A. Bouquillon, J.C. Dran, P. Lagarde, P. Martinetto, B. Moignard, J. Salomon, P. Walter, In-situ dynamic analysis of solids or aqueous solutions undergoing chemical reactions by RBS or PIXE with external beams, *Proc. ECAART 7*, Guildford, in: NIM-B, **188** (2002) pp. 156–161.
[119] E. Ioannidou, D. Bourgarit, T. Calligaro, J.C. Dran, M. Dubus, J. Salomon, P. Walter, RBS and NRA with external beams for archaeometric applications, *Proc. IBA-14*, Dresden, July 1999, in: NIM-B, **161–163** (2000), pp. 730–736.
[120] L. Espie, M. Aucouturier, Surface modification issues in Art, *Surface Engineering*, **17** (2001) 3, 205–216.
[121] M. Aucouturier, Materials science and surface science applied to cultural heritage treasures, in: *Proc. of the 12th Osaka City University International Symposium*, October 27–29, 2004, pp. 17–22.
[122] M. Dubus, I. Colson, J.C. Dran, E. Ioannidou, B. Moignard, J. Salomon, P. Walter, Elemental and structural analysis of corroded lead seals from the XIIIth. to XVth. centuries: preliminary study, in: *Proc. 6th International Conference on Non-Destructive Testing and Microanalysis for the Diagnostics and Conservation of the Cultural and Environmental Heritage*, vol. 2, Rome, May 17–20, 1999, pp. 1741–1750.
[123] J.C. Dran, M. Dubus, B. Moignard, J. Salomon, RBS analysis of thin metallic layers used to monitor museum environment, in: *Proc. Research for Protection, Conservation and Enhancement of Cultural Heritage Opportunities for European Enterprises*, Strasbourg, November 2000.
[124] I. Colson, C. de Grigny, M. Dubus, Les chartes scellées par des bulles de plomb et leur conservation aux Archives Nationales, La Gazette des Archives Nationales, 2002.
[125] M. Dubus, M. Aucouturier, J.C. Dran, O. Guillon, B. Moignard, L. Pichon, J. Salomon, Corrosion monitoring for preventive conservation of the metallic cultural heritage, in: *Proc. ICS Conservation Science 2002*, Edinburgh, May 2002.
[126] M. Dubus, M. Aucouturier, S. Colinart, I. Colson, J.C. Dran, M. Gunn, A.M. Laurent, M. Leroy, R. May, B. Moignard, J. Salomon, P. Walter, Evaluation de la corrosion du plomb aux Archives Nationales de France, in: *Proc. 13th ICOM-CC Triennial Meeting*, Rio de Janeiro, James and James, London, 2002.
[127] M. Dubus, M. Aucouturier, B. Moignard, Atmospheric corrosion monitoring of silver in museums, in: *Proc. Indoor Air Quality*, Norwich, April 28, 2003.
[128] L. Bertrand, T. Calligaro, J.C. Dran, M. Dubus, M.F. Guerra, B. Moignard, L. Pichon, J. Salomon, P. Walter, Développement expérimental d'une ligne PIXE-XRF pour les matériaux du patrimoine, in: *Proc. Congrès Rayons X et Matière 01*, Strasbourg, November 2001, *J. Phys. IV France*, **12** (2002), 359–364.
[129] T. Calligaro, J. Castaing, J.C. Dran, B. Moignard, J.C. Pivin, G.V.R. Prasad, J. Salomon, P. Walter, ERDA with an external helium ion micro-beam: advantages and potential applications, *Nuclear Instruments and Methods* B, **181** (2001), 180–185.
[130] Anonymous, *Nuclear Analytical Techniques in Archaeological Investigations*, Technical reports series N° 416, IAEA, Wien, 2003.
[131] M.F. Guerra, The study of the characterization and provenance of coins and other metalwork using XRF, PIXE and Activation Analysis. in: D.C. Creagh and D.A. Bradley (Eds.), *Radiation in Art and Archaeometry*, Elsevier, Amsterdam, (2000), pp. 417–444.
[132] M.F. Guerra, Ancient gold and modern techniques: identification of gold supplies from Croesus to Mohamed using accelerators and ICP-MS, in: *Non-Destructive Testing and Micro-Analysis for the Diagnostic and Conservation of the Cultural and Environmental Heritage*, *Proc. Art 2002*, Antwerpen, June 2002.
[133] P. Walter, M. Bréniaux, E. Dooryhée, M.A. Lefebvre, P. Martinetto, G. Richard, J. Talabot, G. Tsoucaris, Making make-up in ancient Egypt, *Nature*, 397 (1999), 483–484.
[134] P. Martinetto, M. Anne, E. Dooryhée, G. Tsoucaris, P. Walter, Les formulations cosmétiques de l'Egypte ancienne retrouvées grâce à la diffraction des rayons X, in: *Proc. Congrès Rayons X et Matière 99*, Strasbourg, December 1999, *J. Phys. IV*, **10**, 2000.
[135] L. Bertrand, P. Chevallier, J. Doucet, A. Simionovici, G. Tsoucaris, P. Walter, Apport du rayonnement synchrotron à l'étude des cheveux archéologiques, *J. Phys. IV*, **12** (2002), 359–363.

[136] A.D. Smith, T. Pradell, J. Molera, M. Vendrell, M.A. Marcus, E. Pantos, Micro EXAFS study into the oxidation states of copper coloured Hispano-Moresque lustre decorations. *J. Phys. IV*, **104** (2003), pp. 519–522.

[137] E. Pantos, N. Salvado, T. Pradell, J. Molera, M. Vendrell, A.D. Smith, I.W. Kirkman, M. Marcus, E. Gliozzo and I. Memmi-Turbanti, SR-based molecular speciation of archaeomaterials. in: *Proc. Workshop on Structural Molecular Archaeology*, Ettore Majorana Centre, Erice, Italy, May 25–27, 2002 (in press).

[138] N. Salvadó, T. Pradell, E. Pantos, M.Z. Papiz, J. Molera, M. Seco and M. Vendrell-Saz, Identification of copper based green pigments in Jaume Huguet's Gothic altarpieces by Fourier transform infrared micro-spectroscopy and synchrotron radiation X-ray diffraction, *J. Sync. Rad.*, **9** (2002), 215–222.

[139] E. Pantos, N. Poolton, A. Malins, F. Quinn, J. Prag, Synchrotron-Induced Luminescence from Ceramics and Glazes in the XUV/Soft X-Ray Region, in: *Proc. 4th Symposium on Archaeometry of the Hellenic Society of Archaeometry*, May 28–31, 2003, Athens.

[140] W. Kockelmann, E. Pantos, Synchrotron X-ray and neutron diffraction study of metal artefacts from the Poulton archaeological site in: *Proc. "The use of ion beams in material sciences, medicine and archaeometry"*, Namur, Belgium, May 22–24, 2003 (to be published in NIM B).

[141] T. Pradell, J. Molera, M. Vendrell, J. Perez-Arantequi, E. Pantos, M. Roberts, M. DiMichiel, Role of cinnabar in lustre production, *J. Am. Ceram. Soc.*, **87** (2004) 6, 1018–1023.

[142] M. Müller, M.Z. Papiz, D.T. Clarke, M.A. Roberts, B.M. Murphy, M. Burghammer, C. Riekel, E. Pantos, J. Gunneweg, Identification of ancient textile fibres from Khirbet Qumran caves using synchrotron radiation microbeam diffraction, *Spectrochimica Acta B*, (2004) (in press).

[143] E. Pantos, T. Pradell, N. Salvadó, J. Molera, M. Vendrell, A.D. Smith, I.W. Kirkman, E. Gliozzo, I. Memmi-Turbanti, L. Burgio, G. Martin, W. Kockelmann, A.J.N.W. Prag, Synchrotron radiation and neutrons in archaeological and cultural heritage science, in: *Proc. International School of Physics "Enrico Fermi"*, Varenna, Italy, 2003 (in press).

[144] I. De Ryck, A. Adriaens, F. Adams and E. Pantos, A comparison of microbeam techniques for the analysis of corroded ancient bronze objects, *Analyst*, **128** (2003) 8, 1104–1109.

[145] M. Müller, M.Z. Papiz, D.T. Clarke, M.A. Roberts, B.M. Murphy, M. Burghammer, C. Riekel, E. Pantos and J. Gunneweg Identification of textiles from the Khirbet Qumran caves using microscopy and synchrotron radiation x-ray fibre diffraction, in: J.B. Humbert, and J. Gunneweg (Eds.), *Khirbet Qumran et Ain Feshkha, II. Etudes d'anthropologie, de physique et de chimie*, Fribourg (Suisse), Academic Press, Saint-Paul, 2003.

[146] L. Burgio, R.J.H. Clark, G. Martin, E. Pantos, M.A. Roberts, A multidisciplinary approach to pigment analysis: king's yellow and dragon's blood from the Winsor and Newton pigment box at the Victoria and Albert Museum in Molecular and Structural Archaeology, in: G. Tsoucaris, and J. Lipkowski (Eds.), *NATO Science Series,* Kluger Academic Publishers, 2003, pp. 61–72.

[147] E. Pantos, C.C. Tang, E.J. MacLean, K.C. Cheung, R.W. Strange, P.J. Rizkallah, M.Z. Papiz, S.L. Colston, M.A. Roberts, B.M. Murphy, S.P. Collins, D.T. Clark, M.J. Tobin, M. Zhilin, K. Prag, A.J.N.W. Prag, Applications of synchrotron radiation to archaeological ceramics modern trends in scientific studies on ancient ceramics, in: V. Kilikoglou, A. Hein, Y. Maniatis (Eds.), *BAR International Series 1011*, 2002, pp. 377–384.

[148] W. Kockelmann, E. Pantos, A. Kirfel, Neutron and synchrotron radiation studies of archaeological objects in: D.C. Creagh, D.A. Bradley (Eds.), Radiation in Art and Archaeology, June 2000, Elsevier, Amsterdam.

[149] P. Dillmann, P. Fluzin, P. Chevallier, Determination of iron making processes using synchrotron microprobe, *British Archaeological Reports International Series*, **1043(II)** (2002), 327–344.

[150] I. Reiche, T. Calligaro, M. Menu, J. Salomon, C. Vignaud, Comparative analysis of odontolite, heated fossil ivory and blue fluorapatite by PIXE/PIGE and TEM, *Proc. IBA-14*, Dresden, July 1999, in: NIM-B, **161–163** (2000), pp. 737–742.

[151] I. Reiche, M. Menu, C. Vignaud, Heat induced transformation of fossil mastodon ivory into turquoise "odontolite". Structural and elemental characterisation, *Solid States Sciences*, **2** (2000) 6, 625–636.

[152] H.M. Rietveld, *J. Appl. Cryst.* **2**, (1969), 65–74.

[153] W. Kockelmann, C. Neelmeijer, *Non-destructive fingerprint for the characterisation of Böttger stoneware*, ISIS Report 14022, July 2003.

[154] S. Siano, W. Kockelmann, L. Bartoli, *Archaeometallurgical studies of bronze artefact*, ISIS Report 12605, July 2003.

[155] S. Siano, W. Kockelmann, L. Bartoli, M. Zoppi, M. Iozzo, M. Miccio, Characterisation of archaeological bronze by time of flight neutron diffraction, in: *Proc. 33rd Intern. Symposium on Archaeometry*, Amsterdam, 2004.

[156] W. Kockelmann, A. Kirkel, E. Jansen, R. Linke, M. Schreiner, R. Traum et al., Neutron diffraction for non-destructive texture analysis of minted and cast "Taler" coins, in: *Proc. Numismatics and Technology Questions and Answers*, Kunsthistorisches Museum, Wien, 2003, pp. 113–123.

[157] C. Andreani, G. Gorini, A. Filabozzi, E. Perelli-Cippo, A. Pietropaolo, M. Tarrocchi, Texture and structure studies on marbles from Villa Adriana, in: *Proc. 1st International Workshop on Science, Technology and Cultural Heritage*, Venice, June 29–July 1, 2004, AIV.

[158] B. Mille, D. Bourgarit, L'analyse des alliages anciens à base de cuivre : état des connaissances et développement d'un protocole d'analyse par ICP-AES, Revue d'archéométrie, **24**, 2000, 13–26.

[159] D. Bourgarit, B. Mille, The elemental analysis of ancient copper-based artefacts by inductively-coupled-plasma atomic-emission spectrometry (ICP-AES): an optimized methodology reveals some secrets of the Vix Crater, *Measurement Science and Technology*, **14** (2003), 1538–1555.

[160] L. Simonot, M. Elias, M. Menu, Pigment recognition in works of art thanks to a spectrophotometric database, in: *Colour and Visual Scales 2000*, University Derby, NPL, Egham, UK.

[161] G. Dupuis, M. Elias, M. Menu, L. Simonot, Couleurs, surfaces et pigments au C2RMF, *Techné*, **13–14** (2001) 49–55.

[162] M. Elias, M. Menu, Experimental characterisation of random metallic rough surface by spectrophotometric measurements in the visible range, *Optics communications*, **180** (2000) 191–198.

[163] L. Simonot, A. Zobelli, M. Elias, J. Salomon, J.C. Dran, Microscopic observations and ion beam analyses of pigment distribution in painting glazes, *J. Trace and Microprobe Tech.*, **21** (2003), 35–48.

[164] L. Simonot, M. Elias, Color change due to surface state modification, *Colour Research and Applications,* **28** (2003), 45–49.

[165] G. Dupuis, M. Elias, L. Simonot, Pigment identification by fibre-optics diffuse reflectance spectroscopy, *Applied Spectroscopy*, **56** (2002), 44–51.

[166] M. Elias, L. Simonot, M. Menu, Bidirectional reflectance of a diffuse background covered by a partly absorbing layer, *Optics Communication,* **191** (2001), 1–7.

[167] P. Garside, P. Wyeth, Characterisation of plant fibres by infrared spectroscopy, *Polymer Preprints*, **41** (2000) 2, 1792–1793.

[168] M. Takami, P. Wyeth, Studies on a Korean painted silk banner: identification of layer structure, binding medium and pigments, in: *Preprints of the North American Textile Conservation Conference*, April 2002, pp. 135–144.

[169] P. Garside, P. Wyeth, Characterisation of silk deterioration, in: *Preprints of the North American Textile Conservation Conference,* April 2002, pp. 55–60.

[170] P. Garside, P. Wyeth, Using contemporary analytical techniques to characterise natural fibres in historical artefacts, in: *Proc. 40th Eastern Analytical Symposium*, October 1–4, 2001, Atlantic City, NJ.

[171] S. Pagès-Camagna, S. Colinart, C. Coupry, Fabrication processes of archaeological Egyptian blue and green pigments enlightened by Raman microscopy and scanning electron microscopy, *J. Raman Spectros.*, **30** (1999), 313–317.

[172] S. Pagès-Camagna, S. Colinart, The Egyptian green pigment: its manufacturing process and links to Egyptian blue. Archaeometry, **45** (2003) 637–638.

[173] F. Adar, A. Whitley, B. Roussel, S. Morel, J. Reffner, Innovation in Raman instrumentation for practical applications, in: *Proc. XVIIIth ICORS Budapest*, August 25–30, 2002.

[174] L. Bussotti, E. Castellucci, M. Picollo, Identification of ancient pigments by Raman microscopy, in: *Proc. Congress: Science and Technology for the Safeguard of the Cultural Heritage in the Mediterranean Basin*, vol. I, 1995, pp. 695–699.

[175] R.S. Cataliotti, A. Morresi, A. Romani, N. Forini, L. Bussotti, E. Castellucci, Chemical physical characterization of materials of archaeological and artistic interest. in: *Proc. Congress: Science and Technology for the Safeguard of the Cultural Heritage in the Mediterranean Basin*, vol. I, 1995, pp. 807–810.

[176] B. Wehling, P. Vandenabeele, L. Moens, R. Klockenkämper, A. von Bohlen, G. Van Hooydonk, M. de Reu, Investigation of pigments in medieval manuscripts by Micro Raman-spectroscopy and total reflection X-ray fluorescence spectrometry, *Mikrochimica Acta*, **130** (1999), 253–260.

[177] G. Van Hooydonk, M. de Reu, L. Moens, J. Van Aelst, L. Millis, A TXRF and micro-Raman spectrometric reconstruction of palettes for distinguishing between scriptoria of related mediaeval manuscripts, *European J. Inorganic Chemistry*, **1** (1998), 639–644.

[178] H.G.M. Edwards, F. Rull, P. Vandenabeele, E.M. Newton, L. Moens, Mediaeval pigments in the monastery of San Baudelio, Spain: a Raman spectroscopic analysis, *Applied Spectroscopy*, **55** (2001), 71–76.

[179] M. de Reu, G. Van Hooydonk, P. Vandenabeele, L. Moens, A. von Bohlen, R. Klockenkämper, Remarques méthodologiques à propos de l'analyse chimique des pigments utilisés dans les manuscrits enluminés, *Scriptorium*, **53** (1999), 357–372.

[180] P. Vandenabeele, F. Verpoort, L. Moens; Non-destructive analysis of paintings using FT-Raman spectroscopy with fibre-optics, *J. Raman Spectroscopy*, **32** (2001), 263–269.

[181] P. Vandenabeele, L. Moens, H. Edwards, M. de Reu, G. Van Hooydonk, Identification and classification of natural organic binding media and varnishes by micro-Raman spectroscopy, in: *Proc. 15th World Conference on Non-Destructive Testing*, Rome, 2000.

[182] P. Vandenabeele, L. Moens; Micro-Raman spectroscopy of natural and synthetic indigo samples, *The Analyst*, **128** (2003), 187–193.

[183] P. Vandenabeele, L. Moens, Pigments and binders, a study of artists' materials based on Raman spectroscopy and total-reflection X-ray fluorescence analysis (TXRF), in: R. Van Grieken, K. Janssens, L. Van't dack, G. Meersman (Eds.), *Proc. Art 2002, 7th International Conference on Non-Destructive Testing and Microanalysis for the Diagnostics and Conservation of the Cultural and Environmental Heritage.*

[184] T.E. Jeffries, Laser ablation inductively coupled plasma mass spectrometry, in: K. Janssens (Ed.), *Non-Destructive Micro Analysis of Cultural Heritage Materials*, Elsevier, Amsterdam, 2004.

[185] T.E. Jeffries, S.E. Jackson, H.P. Longerich, Application of a frequency quintupled Nd:YAG source ($\lambda = 213$ nm) for laser ablation inductively coupled plasma mass spectrometric analysis of minerals, *J. Analytical Atomic Spectroscopy*, **13** (1998), 935–940.

[186] V. Lazic, R. Fantoni, F. Colao, A. Santagata, A. Morone, Quantitative laser induced breakdown spectroscopy of ancient marbles and corrections for the variability of plasma parameters and of ablation rate, *J. Anal. At. Spectrom.*, **19** (2004) 4, 429–436.

[187] V. Lazic, F. Colao, R. Fantoni, A. Palucci, V. Spizzichino, I. Borgia, Characterisation of lustre and pigment composition in ancient pottery by laser induced fluorescence and breakdown spectroscopy, *J. Cult. Heritage*, **4** (2003).

[188] P. Maravelaki-Kalaitzaki, D. Anglos, V. Kilikoglou and V. Zafiropulos, Compositional characterization of encrustation on marble with Laser induced breakdown spectroscopy, *Spectrochimica Acta B*, **56** (2001), 887–903.

[189] B. Blümich, V. Anferov, S. Anferova, M. Klein, R. Fechete, An NMR-Mouse for analysis of thin objects, *Macromol. Mater. Eng.*, **288** (2003), 312–317.

[190] B. Blümich, S. Anferova, S. Sharma, A.L. Segre, C. Federici, Degradation of historical paper: non-destructive analysis by the NMR-Mouse, *J. Magn. Reson.*, **161** (2003), 204–209.

[191] J. Perlo, F. Casanova, B. Blümich, 3D imaging with a single sided sensor: an open tomography, *J. Magn. Reson.*, **166** (2004), 228–235.

[192] M. Regert, Productions techniques et résidus alimentaires préhistoriques : les apports de la chimie organique, *Techné*, **13–14**, (2001), 71–77.

[193] M. Regert, S.N. Dudd, P. Pétrequin, R.P. Evershed, Fonction des céramiques et alimentation au Néolithique final sur les sites de Chalain: de nouvelles voies d'étude fondées sur l'analyse chimique des résidus organiques conservés dans les poteries, *Revue d'Archéométrie*, **23**, (1999), 91–99.

[194] M. Regert, Investigating the history of prehistoric glues through gas chromatography – mass spectrometry. *J. Separation Science*, **27** (2004), 244–254.

[195] M. Regert, N. Garnier, O. Decavallas, C. Cren-Olivé, C. Rolando, Structural characterization of lipid constituents from natural substances preserved in archaeological environments, *Measurement Science and Technology*, **14** (2003), 1620–1630.

[196] M. Regert, S. Vacher, C. Moulherat, O. Decavallas, Adhesive production and pottery function during the Iron Age at the site of Grand Aunay (Sarthe, France), *Archaeometry*, **45** (2003) 101–120.

[197] M. Regert, C. Rolando, Identification of Archaeological Adhesives Using Direct Inlet Electron Ionization Mass Spectrometry, *Analytical Chemistry*, **74** (2002), 965–975.

[198] N. Garnier, C. Cren-Olivé, C. Rolando, M. Regert, Characterization of Archaeological Beeswax by Electron Ionization and Electrospray Ionization Mass Spectrometry, *Analytical Chemistry*, **74** (2002) 4868–4877.

[199] N. Garnier, *Analyse structurale de matériaux organiques conservés dans des céramiques antiques*, Thesis, Université de Paris 6, 26 février 2003.

[200] N. Eastaugh, V. Walsh, R. Siddall, T. Chaplin, *Pigment Compendium: A Dictionary of Historical Pigments*, Butterworth-Heinemann, London, 2004.

[201] N. Eastaugh, V. Walsh, R. Siddall, T. Chaplin, *Pigment Compendium: Optical Microscopy of Historical Pigments*, Butterworth-Heinemann, London, 2004.

[202] R.G. Roberts, Luminescence dating in archaeology: from origin to optical, *Radiation Measurements*, **27** (1997), 819–892.

[203] J. Castaing, T. Borel, A. Bouquillon, E. Porto, A. Zink, Influence of X-ray induced electronic defects on luminescence dating of works of art, in: *Proc. LED 2002*, Reno, USA, June 24–28, 2002.

[204] T. Higham, F. Petchey, Radiocarbon dating in archaeology; methods and applications. in: D.C. Creagh, D.A. Bradley (Eds.), *Radiation in Art and Archaeometry*, Elsevier, Amsterdam, 2000, pp. 255–285.

[205] C. Tuniz, M. Zoppi, M. Barbetti, AMS dating in archaeology, history, and art. in: D.C. Creagh, and D.A, Bradley (Eds.), *Radiation in Art and Archaeometry*, Elsevier, Amsterdam, 2000, pp. 444–472.

Chapter 2

# X-ray and Neutron Digital Radiography and Computed Tomography for Cultural Heritage

Franco Casali

*Department of Physics, University of Bologna, Italy*
*Email: franco.casali@bo.infn.it*
*www.xraytomography.com*

**Abstract**

Methods of diagnosis based on digital radiography (DR) and computed tomography (CT), are more and more frequently used in the cultural heritage field. The application of these techniques can help restoration and conservation planners to understand historical construction techniques and to reveal poor restoration work and forgeries.

As the size of objects of cultural interest varies greatly, from small fragments (for which high spatial resolution is needed) to large works of art (for which large detectors are necessary), it would not be appropriate to describe any one particular measuring device in detail. In this chapter, we will therefore provide an overview on Digital Radiography (DR) and Computed Tomography (CT) systems, underlining their range of applications. The chapter focuses mainly on X-ray radiation (with different kinds of sources) although neutron DR and CT are also mentioned, as neutron imaging should be considered complementary to X-ray imaging.

Some DR and CT images, most of which were taken by researchers at the Department of Physics of the University of Bologna, are shown. This overview adopts a tutorial approach, as it is aimed at those with no specific knowledge of digital imaging. Three appendices have also been included (concerning Fourier transforms, modulation transfer function and DR and CT acquisition systems) for those readers who wish to acquire further skills in the field of digital imaging.

**Keywords:** X-ray digital radiography, tomography, neutron imaging, neutron CT, cultural heritage digital imaging.

**Contents**

# 1. INTRODUCTION

Physical methods of diagnosis are finding more and more applications in the cultural heritage field either for scientific investigation or for restoration and conservation purposes. It is often vitally important to gain information on the invisible parts of a work of art or archaeological find, revealing the artist's preparatory sketch or changes of idea or for example, to examine the state of corrosion of a bronze statue or the cracks in a marble statue. For additional information, please refer to Chapter 1. These diagnostic techniques refer primarily to the investigations performed using electromagnetic radiation at various wavelengths, particle beams and sound waves. These techniques must be as non-invasive as possible.

## 1.1. Electromagnetic radiation for internal investigations

Several potential applications exist, depending on the wavelength of the electromagnetic radiation:

- wavelength of approximately 10 mm (georadar); (in the cultural heritage field, this technique is used for detecting old foundations, empty rooms, invisible galleries and so on);
- wavelength of approximately 1000–3000 nm (infrared) (useful for detecting preparatory sketches hidden below layers of paint);
- wavelength of less than 5 nm for X-rays and less than 0.1 nm for gamma rays (used for performing radiographs and tomographs).

### 1.2. Particle beams

In the cultural heritage field, neutrons are most frequently used for conducting internal investigations and they have complementary characteristics to those of X-rays and gamma rays. For "surface" or "shallow" investigations, electrons (SEM–TEM) and alpha particles and protons (particle induced X-ray emission = PIXE) can also be used.

### 1.3. Ultrasound and Sonic waves

Sonic waves and ultrasound techniques are mechanical weak impact stresses useful in cases where X-rays have low penetration.

- Sonic waves provide useful information on the interior of brick-built columns or columns built using non-homogeneous materials.
- Ultrasound provides information on cracks or discontinuities in metals or stone objects (*e.g.* columns), where the X-rays are not suitable given the thickness of the object being examined.

Generally speaking, a diagnostic imaging system consists of:

- a radiation source;
- a radiation detector;
- equipment for moving the object in relation to the source–detector loading;
- a computer for managing the image acquisition process;
- a computer for image processing and rendering.

In this chapter, some basic elements on digital radiography (DR) and computed tomography (CT) will be given, relating primarily to X- and $\gamma$ rays (and, to a lesser extent, neutrons). Even if DR is the natural extension of radiography, there is an increasing interest in CT. In fact, this kind of diagnosis gives more information than DR as it is apparent from Figs. 1 to 3.

Figure 1 reproduces the image of a small clay bust, a copy of one found at Pompeii (this image will be our reference for subsequent elaborations). Figure 2 reproduces some radiographies of this small bust and Fig. 3 gives the 3D representation of it after a CT. A small defect, of the order of 800 μm not visible by DR, is clearly detectable by CT (Fig. 3 – right).

This overview is addressed mainly to people working in the field of cultural heritage like restorers, conservators and art critics.

## 2. RADIATION SOURCES

### 2.1. X-rays and $\gamma$ rays

#### *2.1.1. What are X-rays?*

The X- and $\gamma$ rays are produced by "photons", electromagnetic wave packets that can behave either as waves or particles. Photons are characterised by their wavelength, $\lambda$, and

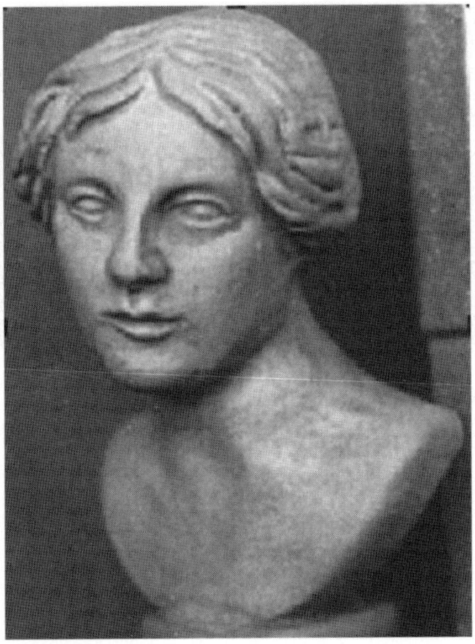

**Fig. 1.** Photo of a small clay head, copy of a find from Pompeii.

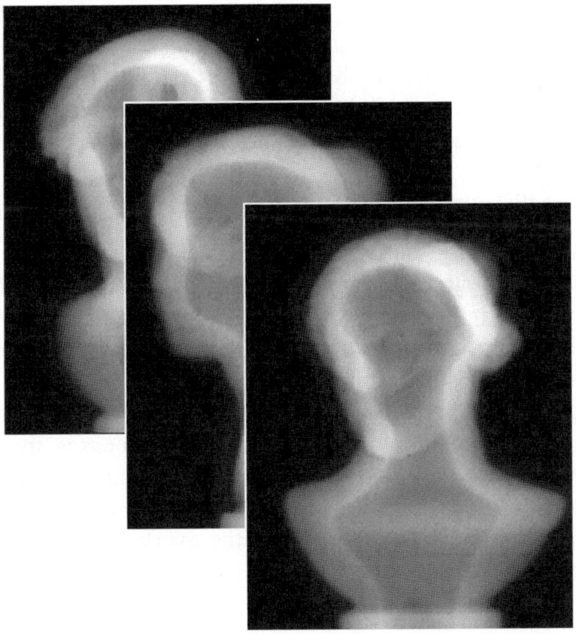

**Fig. 2.** X-ray radiographies of the object in Fig. 1.

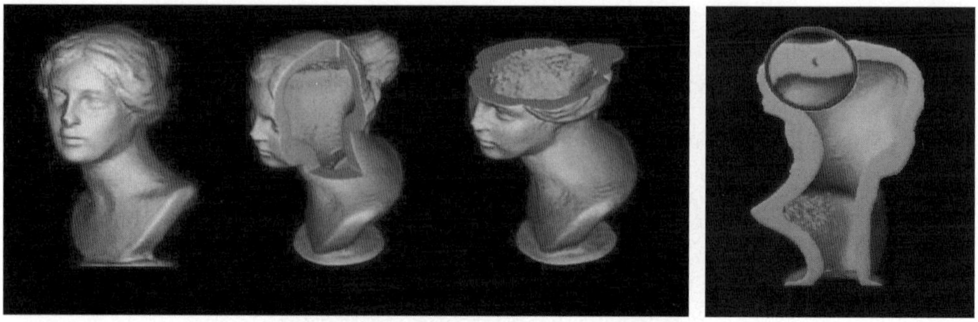

**Fig. 3.** Three-dimensional representation of the 3D CT image of the object in Fig. 1 (left) and a small defect in the clay (about 0.8 mm), not visible on X-ray radiography (right).

energy, $E$. These two quantities are related by the following expression:

$$E = h\nu = hc/\lambda \tag{1}$$

where:

$h$ = Planck constant;
$\nu$ = frequency;
$c$ = light speed.

The wave aspect of photons appears mainly at low frequencies (*e.g.* radio or TV waves), the particle aspect is predominant for high frequencies (high energy). For the applications in question, we can consider that photons act as particles.

The X-ray photons of interest to us have energies ranging from a few keV to several MeV (far more energetic than the visible light photons). As the minimum energy required for ionisation is 10 eV (UV radiation), X-rays are considered as "ionising radiation", and they must be used with care.

X-rays, which were discovered in 1895 by W. Roentgen,[1] can be generated in two ways.

1. When fast electrons undergo acceleration (or deceleration) they emit photons, also known as "bremsstrahlung" radiation. The energy distribution of photons or "energy spectrum" is of the continuous type ("white radiation");
2. When electrons are removed from the innermost orbits of an atom (see Fig. 4), the electrons, which belong to the outer orbits, jump into the "holes" created. In this process of rearrangement, photons are emitted with energies equivalent to the difference between the binding energies of the inner and outer orbits. In such cases, the energies spectrum of emitted photons is a "line spectrum". These particular energies (characteristic of each element) are also known as "lines of X-ray fluorescence". If fast electrons impinge on a material, the resulting X-ray spectrum is the overlap between the continuous and fluorescence spectrum (see Fig. 5).

[1] For his discovery, W. Roentgen received the Nobel Prize for Physics. Roentgen did not accept the prize money: rather he used it to set up fellowships for the best young German physicists.

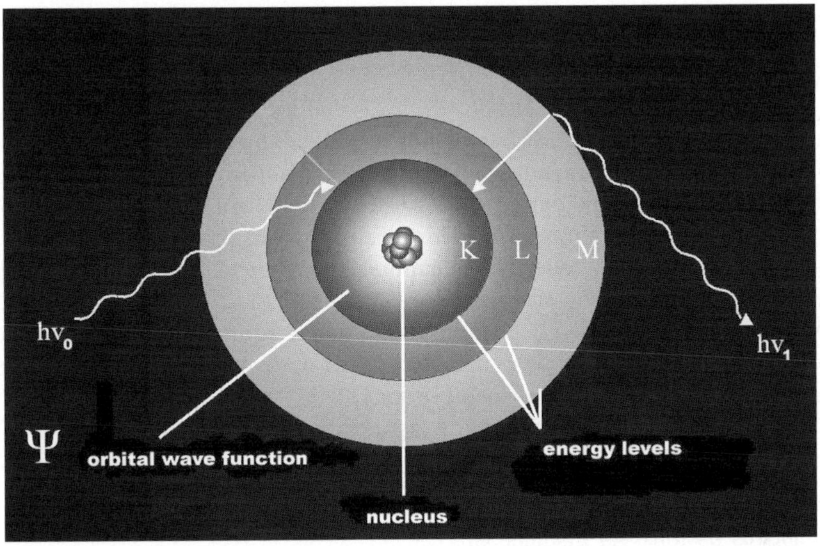


**Fig. 4.** Diagram of atomic shell structure and interaction with ionising photons.

### *2.1.2. What are γ rays?*

Gamma rays are photons. They cannot be distinguished from X-ray photons since both are electromagnetic radiation. Gamma rays are produced during the reassembly of the nucleus after specific nuclear reactions (*e.g.* α or β decays and the capture of other particles). For particular nuclei and particular reactions, the emitted photons always have the same energy and the γ spectrum is of a "line" type. It is possible to identify the radioactive isotopes from the line distribution in the spectrum.

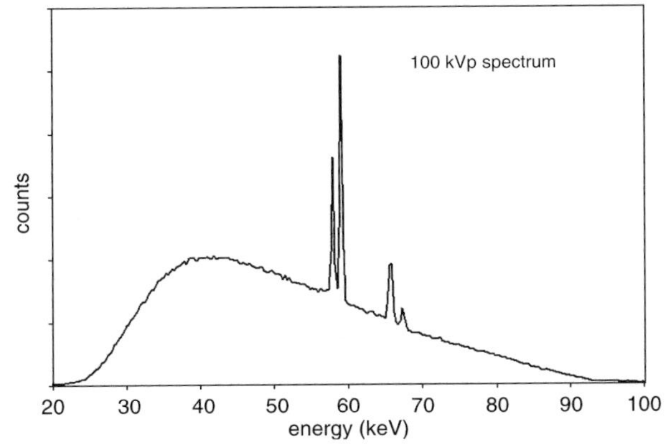


**Fig. 5.** Radiation spectra of a conventional X-ray tube. The continuous component, due to bremsstrahlung, and the discrete component, due to characteristic emission, are visible.

## 2.2. Neutrons

Neutrons, discovered by Chadwick[2] in 1932 are neutral particles that make up part of atomic nuclei. In the free state, neutrons decay as "proton + electron + antineutrino" with a mean life of 1000 s. Having no charge, they penetrate objects easily, which makes them good probes for diagnostic imaging purposes. When neutrons are captured by nuclei, the nuclei become radioactive and emit $\gamma$ rays. By analysing the emitted spectrum, it is possible to infer what the activated elements are. This type of analysis, known as Neutron Activation Analysis (NAA), is several orders of magnitude more sensitive than standard chemical analysis.

As NAA is a non-destructive technique, it is often used in archaeometry for detecting traces of materials (*e.g.* impurities characteristic of materials from a certain mine, thus enabling the identification of the place of origin) [1–3].

Neutrons can be produced by the following means:

1. nuclear reactors, through the fission induced in particular isotopes such as $U^{235}$, $U^{238}$ or $Pu^{239}$;
2. spontaneous fissions, for instance $Cf^{252}$;
3. particular nuclear reactions making use of particle accelerators (for instance, bombarding $Be^9$ with $\alpha$–particles);
4. small accelerators in which the reaction $H^2 + H^3 \rightarrow He^4 + n$ occurs.

## 2.3. X-ray sources

The X-ray sources of interest in this chapter can be summarised as:

- X-ray tubes (from 5 to 450 kV);
- linear accelerators (from 2 to 15 MV);
- synchrotron light (from 5 to 100 keV).

### *2.3.1. X-ray tubes*

A schematic diagram of a typical X-ray tube is shown in Fig. 6. The electrons, produced by a heated filament inside a glass tube – where high vacuum has been created – are accelerated against a target (anode). For electron energy less than 1 MeV, bremsstrahlung radiation is produced mainly perpendicular to the electrons' direction of flight; otherwise, for energy higher than 1 MeV, X-ray radiation is mainly produced in a forward direction [4].

Only a small fraction of the kinetic energy of the electrons is transformed into X-rays: the remainder heats the anode. For good anode cooling, a rotating target (or a cooling circuit) is used, mainly for tubes with powers higher than 100 W. When the object being tested is made of heavy material, industrial tubes are used. These are designed to operate continuously; long exposure time (several hours) is normal. On the contrary, medical tubes are designed to give short high power shots, in order to minimise motion artefacts.

[2] For his discovery, Edwin Chadwick received the Nobel Price for Physics in 1935.

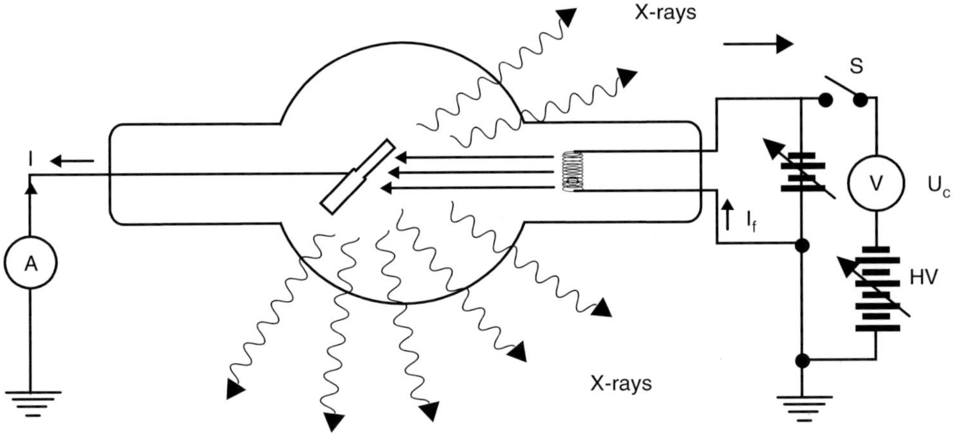


**Fig. 6.** Diagram of an X-ray tube.

The effective size of the anode from which the X-ray beam is emitted is called the "focal spot". Focal spot dimension is very important for image definition. The smaller the focal spot, the sharper is the "shadow" produced by the X-ray beam on the detector. For extended focal spots there is a penumbra known as "source unsharpness", as shown in Fig. 7.

The penumbra dimension can be calculated using the following formula:

$$P = \frac{fD}{d}. \tag{2}$$

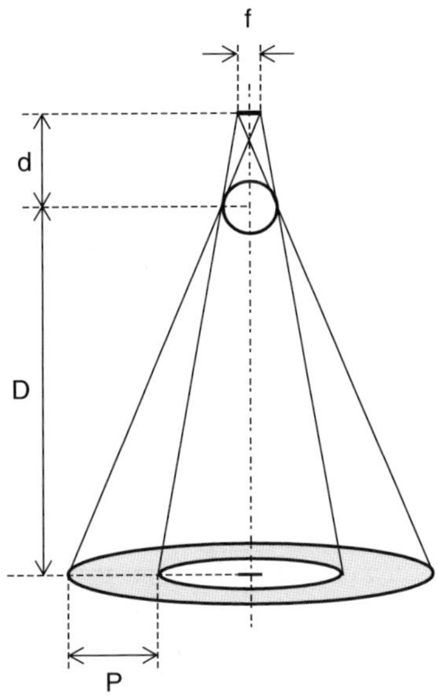


**Fig. 7.** Unsharpness, due to the real size of the focal spot of an X-ray source.

where:

$P$ = width of the penumbra;
$f$ = effective focal spot size;
$D$ = distance between object and image plane;
$d$ = distance between source target and object.

**Microfocus and Nanofocus Tubes.** For having high spatial resolution (low "penumbra"), X-ray tubes, called microfocus tubes, are used where the focal spot is of the order of few microns. X-ray tubes are now available with focal spots that can reach dimensions of 0.5 microns (nanofocus). Because of heat-loading effects in the anode in these (rather expensive) tubes, current is low (a few μA) and maximum voltage does not exceed 150 kV. Using microfocus, or nanofocus, it is possible to obtain the CT of small objects with high spatial resolution.

**Industrial Tubes.** For high currents and voltage up to 450 kV, industrial type tubes are used. Usually they have a cooled anode and current can reach several millamperes. These tubes can be used for radiography or CT of bronze statues of several millimetre thickness. Both microfocuses or industrial tubes operate in a continuous way.

### *2.3.2. Linear accelerators (LINAC)*

Figure 8 shows the scheme of a linear accelerator or LINAC (LINear ACcelerator). The electrons emitted from the cathode are "packaged" and accelerated against the anode by an electromagnetic wave of a suitable frequency (radio frequency), like a surfer carried to the shore by a wave. The derived bremsstrahlung is, therefore, of pulsed type. The pulse frequency can range up to several MHz. The maximum energy of the photons produced is the maximum energy achieved by the electrons; however very few photons have maximum energy. The energy spectrum is continuous (Fig. 9). Without suitable absorbers (filters), one can

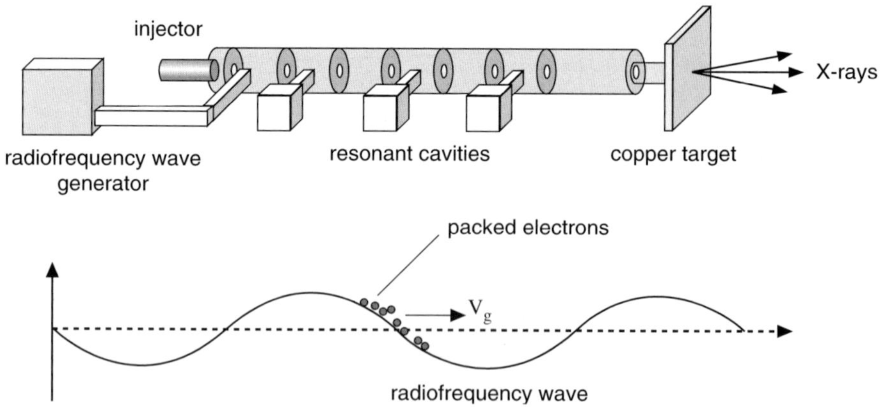


**Fig. 8.** Schematic representation of a linear accelerator with resonant cavities and a copper target struck by accelerated electrons. High-energy photons are produced by this interaction.

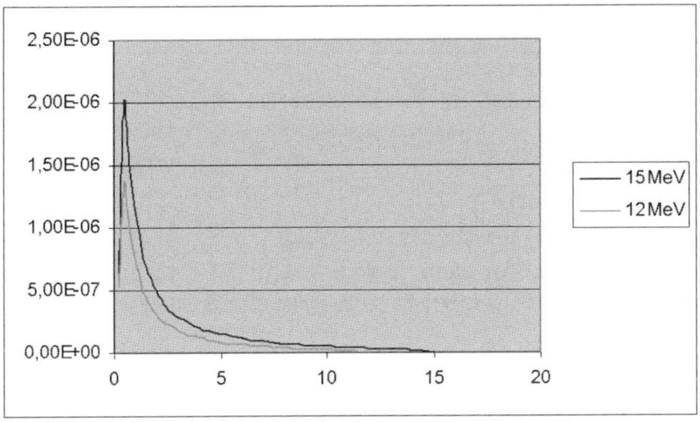


**Fig. 9.** X-ray spectra produced by a 12 MeV and 15 MeV LINAC (courtesy Dan Schneberk, LLNL).

assume the "effective energy" (equivalent to a monochromatic source) to be one-third of the maximum energy or less.

Linear accelerators can be used for DR or CT of thick or high-density objects (for instance, see Ref. [5]). Portable LINACs do exist, however to the author's knowledge, they are not yet used "on the field" for cultural heritage applications.

*2.3.3. Synchrotrons*

Synchrotrons are electron accelerators shaped like a large ring. Electrons can achieve energies of several GeV. If electrons are compelled to move out of their orbit by deflecting magnets or by arrays of bending magnets, they emit an X-ray radiation named "synchrotron light". This radiation, ranging from 5 to 100 keV, can be selected in energy by proper monochromator crystals, making use of "Bragg's law". The synchrotron light is so intense that it is possible to obtain very high energy definition [6–8].

### 2.4. Radioisotope sources

At present, the radioisotope sources most commonly used in the cultural heritage field for the radiographic analysis of statues and other works have been $^{60}Co$ and $^{137}Cs$, as in the case of the radiographs performed on the arm of Michelangelo's David [9], the Riace bronzes, a Roman bronze statue [10], and so on.

The *advantages* of using isotopic sources are their low cost (in comparison to the LINAC), the single energy of the emitted photons, and the small dimension of the probes, which enables inspections that would otherwise be impossible.

The *disadvantages* are: the source dimensions (corresponding to a large focal spot), the difficulty of transportation (due to shielding and safety limitations) and handling, and their decrease in intensity (rather low) with time.

**Table 1.** Important characteristics of two radioisotopic sources (from Ref. [11])

| Isotope | $^{60}Co$ | $^{137}Cs$ |
|---|---|---|
| Half-life (year) | 5.3 | 30 |
| Gamma ray(s) energy (MeV) | 1.17 and 1.33 | 0.66 |
| Practical source diameter (mm) | 3 | 10 |
| Al half-value thickness* (mm) for each γ ray listed | 42 and 48 | 34 |
| Fe half-value thickness* (mm) for each γ ray listed | 15 and 17 | 12 |

* The "half-value thickness" is the thickness of a material that reduces the beam intensity to half.

The disintegration intensity decreases exponentially over time:

$$n(t) = n_0 e^{-\lambda t} \tag{3}$$

where:

$n(t)$ = disintegration number at the time $t$
$n_0$ = disintegration number at the time $t = 0$, when the intensity of the source is defined
$\lambda$ = decay rate (disintegration/second).

The quantity $\lambda$ is related to the half-time, $T^{1/2}$, by the following relation:

$$\lambda = \frac{\ln 2}{T^{1/2}} = \frac{0.693}{T^{1/2}}. \tag{4}$$

The source power, that is the disintegration rate (dis/s), is expressed in "becquerel" (Bq) (one Bq corresponds to one disintegration or transmutation per second). In the past, "curie" (Ci), corresponding to $3.7 \times 10^{10}$ dis/s, was used as a unit of measurement.

Table 1 gives the characteristics of $^{60}Co$ and $^{137}Cs$, the two commonly used isotopic sources.

## 3. INTERACTION OF THE RADIATION WITH MATTER

### 3.1. General considerations

The imaging diagnostic techniques in question (radiography and tomography) concern the attenuation of particle beams through their interaction with matter. We consider both neutrons and photons as particles.

In "good geometry" conditions, for parallel particle beams, the attenuation follows the Beer-Lambert's Law:

$$I(d) = I_0 e^{-kd} \tag{5}$$

where:

$I(d)$ = the particle number passing through a body with a thickness $d$;
$I_0$ = the particle number which reaches the detector without the body;
$k$ = radiation attenuation coefficient.

For photons, $k$ is indicated by the Greek letter $\mu$ (linear attenuation coefficient ($cm^{-1}$)) or by $\mu/\rho$ (mass attenuation coefficient ($cm^2/g$)), where $\rho$ is density ($g/cm^3$). In this case, equation (5) becomes:

$$I(d) = I_0 e^{-\mu d} \tag{6a}$$

or

$$I(d) = I_0 e^{-(\mu/\rho)\rho d}. \tag{6b}$$

For neutrons, $k$ is indicated by the Greek letter $\Sigma$ (total macroscopic cross section ($cm^{-1}$)) and equation (5) becomes:

$$I(d) = I_0 e^{-\Sigma d}. \tag{7}$$

Both $\mu$ and $\Sigma$ are rather complicated functions which depend on the irradiated material and particle energy. Figures 10 and 11 show typical shapes of ($\mu/\rho$) and $\Sigma$.

### 3.2. Good geometry

In physical measurements, "good geometry" is used to describe a situation in which a particle, which interacts with the medium under investigation – so that is removed from the beam – does not interact in any other way with the detector. This does not occur when, after one or more shots, the particle is deviated on the detector and is counted as though it has had no interactions with matter. Figure 12 clarifies this concept.

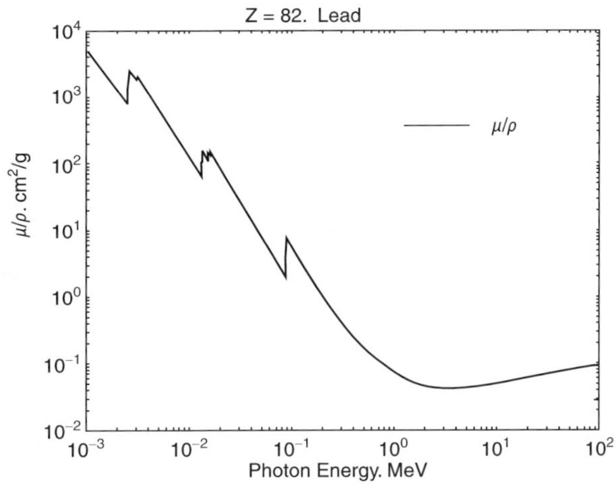


**Fig. 10.** X-ray mass attenuation coefficient for lead (From Ref. [50]).

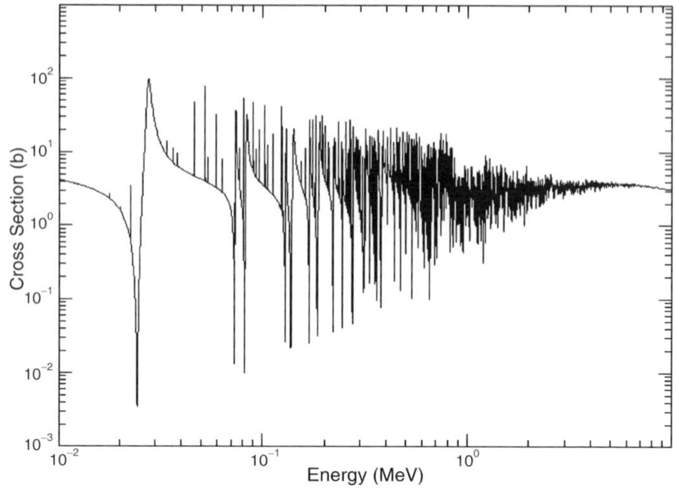


**Fig. 11.** Neutron microscopic total cross section.

This indirect component is called "diffused radiation"; sometimes in poor-geometry conditions, the diffused radiation is one order of magnitude larger than the direct radiation. Diffused radiation can be eliminated, or decreased, by the use of suitable collimators (see Fig. 13).

### 3.3. "Beam hardening" for photons

Usually the smaller the energy of interacting particle, the higher will be the attenuation coefficients. When the radiation is not monochromatic (Fig. 5), the weaker component is

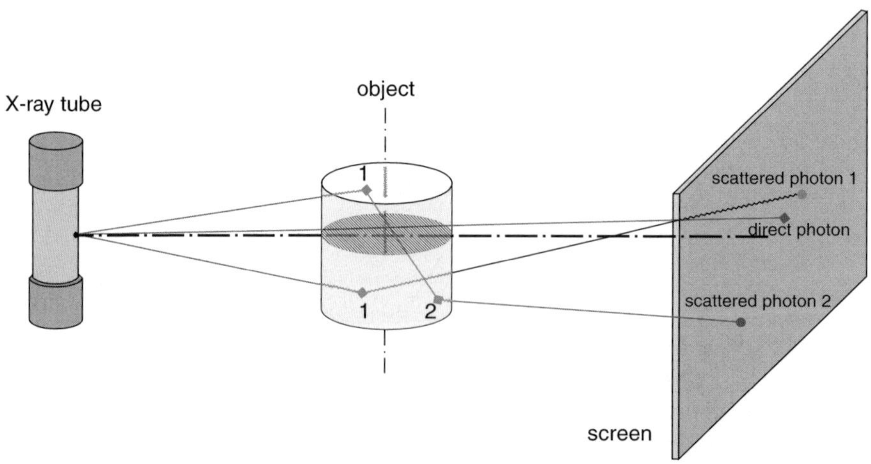


**Fig. 12.** Interaction of photons (scattered and unscattered) with an object.

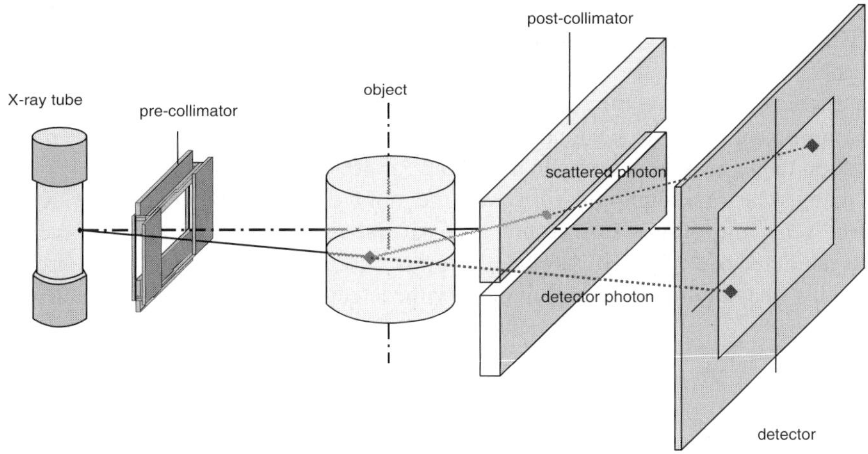


**Fig. 13.** Pre-collimation and post-collimation of the radiation beam to reduce the scattered component.

absorbed more easily than the harder one; consequentially, when the radiation penetrates deep into the object, the energetic spectrum becomes harder and harder. This phenomenon is called the "beam hardening effect". In CT, where the absorption coefficient is assumed to be constant with energy, it is necessary to correct this "hardening", which is equivalent to a variation of $\mu$ inside the object, even if it has a homogeneous composition.

# 4. DIGITAL IMAGING FOR X- AND $\gamma$ RAYS

## 4.1. General considerations

In an ideal detection system, a photon, originating from a point source, reaches the detector with a probability given by Beer's law, expressed by equation (6a). However, a large number of factors render this equation invalid. Firstly, it should be pointed out that the radiation source is not a point (*e.g.* focal spot of finite dimension for X-ray tubes). The second reason is the photon diffusion over the detector during interaction with the object and the experimental fixtures (collimators, room, walls, etc.). The third cause is the photon diffusion inside the detector. The image degradation created by these three causes is known as "blurring". If $N_p$ is the number of primary photons, which arrive at the detector, and $N_s$ is the number of scattered photons, the ratio ($N_p/N_s$) can be considered as the ratio between the true signal ($N_p$) and noise ($N_s$). A simple increase in primary radiation does not increase the "signal-to-noise-ratio". This can be achieved using suitable collimators, adjusting the object–screen distance, decreasing the detector thickness, and so on. One very important characteristic of a detection system is the "dynamic range", defined as the ratio of the maximum and minimum detectable signal. If we consider radiographic film, it suffers from an intrinsic noise ("fog") and shows a maximum exposition level beyond which there is "saturation" (all the silver grains are separated from the iodine). The dynamic range is stated in "decades" or in "bits" (see Section 4.2.3). When we say that a film has a dynamic range of 3 decades, we mean that the radiation intensity which gives saturation is 1000 times the intrinsic noise; and when we say that a detector has a dynamic range of 12 bits, we mean that its operation range goes from intrinsic noise to $2^{12} = 4096$ times the background.

For many years radiographic film was the only detector and register of X- and γ radiation. Very fine grain films assure high spatial resolution [12] and are considered "analogical detectors". Today "digital detectors" (see Section 5) are more frequently used. In this chapter, we will deal mainly with detectors suitable for the acquisition of digital images.

## 4.2. Image digitising

### *4.2.1. Foreword*

Let us assume a wish to "digitise" an image acquired by a radiographic film. Such an operation can be performed by: (a) taking the film and reading the degree of transparency (related to the "optical density") by a small detector by moving it step by step and (b) transforming the obtained value into a binary one by means of an Analogue to Digital Converter (ADC).

The "digital" image obtained is a matrix of numbers, similar to a chessboard. The procedure described is commonly used in scanners for transforming "analogue images" (photos or texts) into "digital images". The smallest matrix element is called a *pixel* (PICture ELement) (Fig. 14(A)). Therefore, we are in the presence of two types of discretisation: the first concerning spatial sampling, the second, the subdivision of the grey interval (from black to white). A digitising scheme for an analogue signal is given in Fig. 14(B). If we normalise the intensity range, associating the black to zero and the white to one, the problem is how to pass from the continuous grey interval (black → white) into a finite number of grey levels.

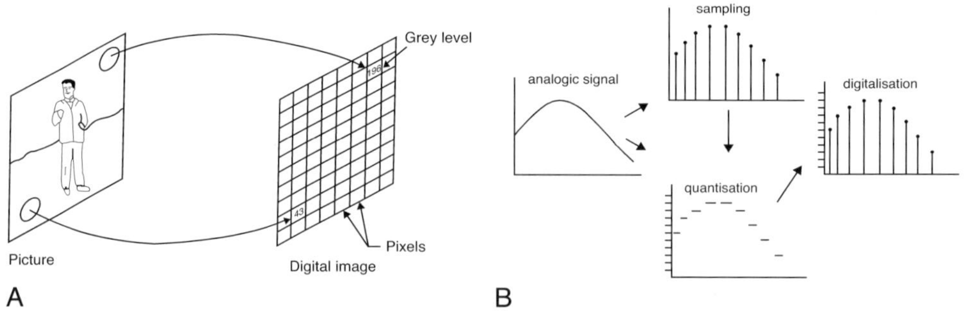


**Fig. 14.** (A) Image digitising scheme; (B) Analogue signal digitising scheme.

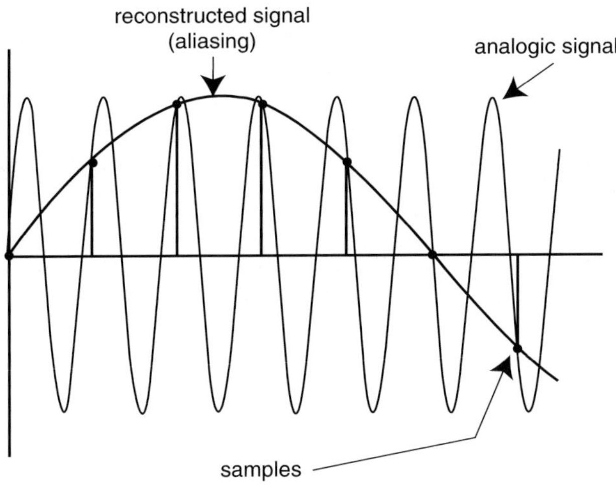


**Fig. 15.** A periodic signal is sampled with a pitch larger than $\lambda/4$ and the reconstructed signal is different from the original.

### *4.2.2. The "sampling theorem" for spatial reproduction*

*Spatial resolution*, that is the quality of the reproduction, depends on the "sampling pitch". A theorem[3] (called "Nyquist or Shannon sampling theorem") says that, unambiguous imaging of a feature of size *d* is best performed when the sampling pitch is less then *d*/2. Inadequate sampling results in detail loss: in such cases, we obtain so-called "aliasing"[4]. An example of aliasing is shown in Fig. 15. In this example, a periodic signal is sampled with a pitch larger than $\lambda/4$, an amount larger than that allowable by the Nyquist theorem, and the reconstructed signal is totally different ("alias") from the original one.

Figure 16 reproduces the image of the clay bust represented in Fig. 1, at different sampling pitches (256, 64, 32, 16 dots per inch (dpi)). Obviously many details are lost when a very large sampling pitch is used.

### *4.2.3. Discretising the grey interval*

Having "sampled" the image in space, for each pixel we must allocate a "number" to the grey level. Once again, the quality of the reproduction of the grey tones depends on how many sub-intervals the "black-white" range is subdivided into. For colour images, one has to discretise each primary colour (red, green, blue). If the sub-intervals are too few, *e.g.* $16 = 2^4$ (ADC at 4 bits), reproduction will be coarse; if the number is high, *e.g.* $256 = 2^8$ grey levels

[3]The sampling theorem was stated by Nyquist in 1928 and mathematically proven by Shannon in 1949. This sampling theorem is called "Nyquist Sampling Theorem", or "Shannon Sampling Theorem" and it is valid in the acoustic field too.

[4]A terrible word obtained declining in English the Latin word "alias"!

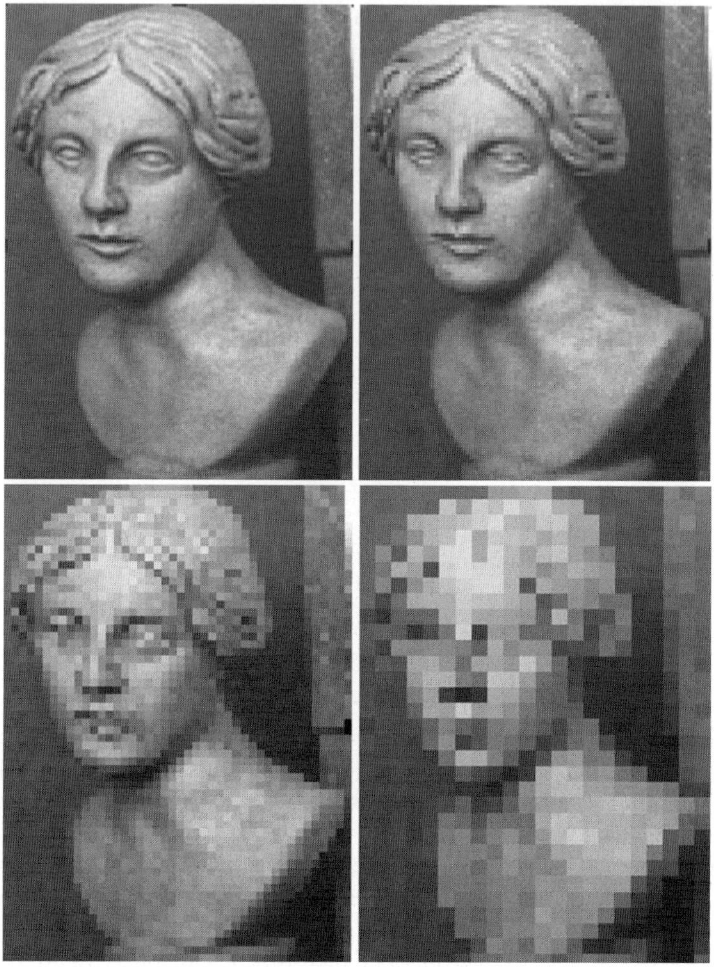

**Fig. 16.** The test digital image as a function of sampling pitch. The aliasing effect is evident for low pixel numbers (large sampling pitches).

(ADC at 8 bits), reproduction may be acceptable. As the human eye cannot distinguish more then 15–20 shades of grey, 8 bits (1 byte) are usually sufficient to give good reproduction for standard photos. However, for digital radiographs of objects of interest in cultural heritage, (a bronze statue, for instance) 8 bits are insufficient as we have to discriminate between very close grey levels. Modern digital systems for DR can use ADC up to 14 bits and more. If we use only 2 grey levels (black and white), that is 1 bit, we will obtain a bit-map. Figure 17 shows images of the same small bust taken with decreasing bits (8, 4, 2, 1).

In the past, many mathematical techniques have been developed to improve the quality of digital images. These techniques (which will be dealt with briefly later) are related to enhancement and to the more complex field of image restoration.

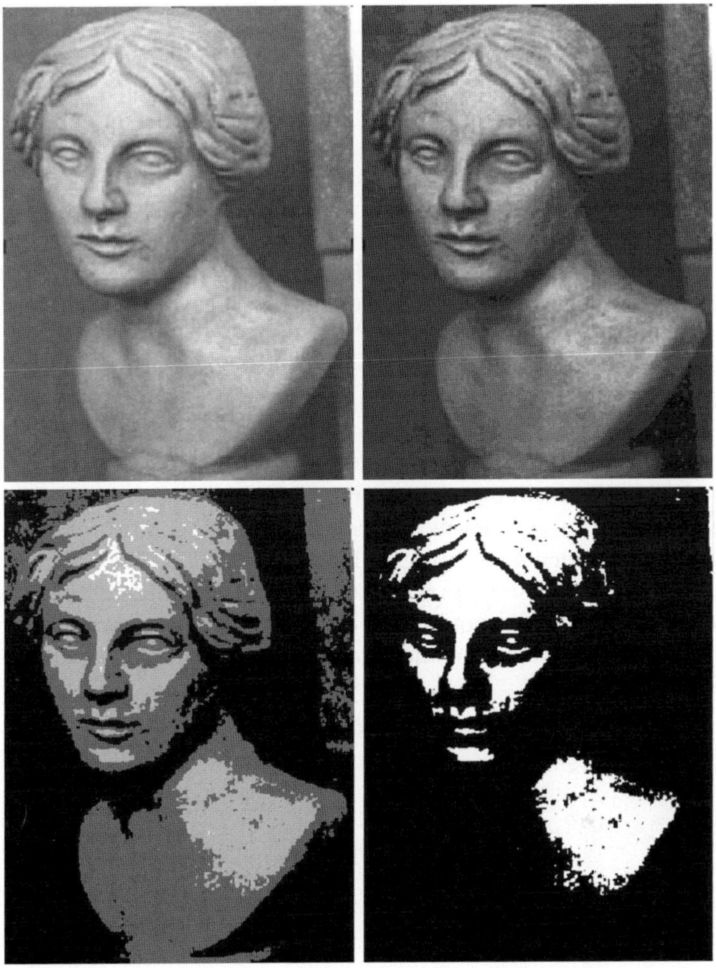

**Fig. 17.** The test digital image as a function of the number of bits.

## 4.3. Image enhancement

### *4.3.1. The histogram of a digital image*

Let us suppose we have acquired an image "*f*" by a planar detector with $M$ rows and $N$ columns at 8-bit grey level. This means that our image is equivalent to a matrix of $N \times M$ numbers ranging from 0 to 255. We can now count how many pixels have a grey level, $r_k$, and then create the discrete function $p_f(r_k)$, named the "histogram of the digital image", defined for 256 grey levels only.

Sometimes the histogram is given in a "normalised" form:

$$pf(rk) = nk/n \tag{8}$$

where:

$k = 0, 1, 2, \ldots, 255$;
$n_k$ = how many times the $k$-level appears in the image, that is the number of pixels having the grey level $r_k$;
$n$ = total number of pixels ($M \times N$).

In this definition, $p_f(r_k)$ is a function with values between 0 and 1 and the integral equal to 1. Sometimes $r$ is also normalised in the range between 0 (black) and 1 (white). Figure 18 shows the histogram of the image shown in Fig. 1.

Two different images can have the same histogram, as illustrated in Fig. 19. All histograms shown were obtained by "Adobe Photoshop" software.

For a bit map, the histogram is made by 2 segments, one that gives black pixels, the other that gives the white ones. Histograms can also be taken of colour images (one for each channel: R, G, B).

From the shape of the histogram it is possible to infer characteristics of the image. A grey image, with poor contrast, will have a histogram similar to that shown in the left image of Fig. 20 (predominance of grey levels). In contrast, a high contrast image, on the right, shows two peaks corresponding to two grey levels (0 and 255). The object is well-distinct from the background. This histogram is said to be "bimodal". Figure 21 refers to dark and bright images respectively. It is possible to operate on the histogram, pixel by pixel, by substituting a grey level, $r$, with another grey level, $s$, where $s$ is obtained from $r$ by a transformation law:

$$s = T(r). \tag{9}$$

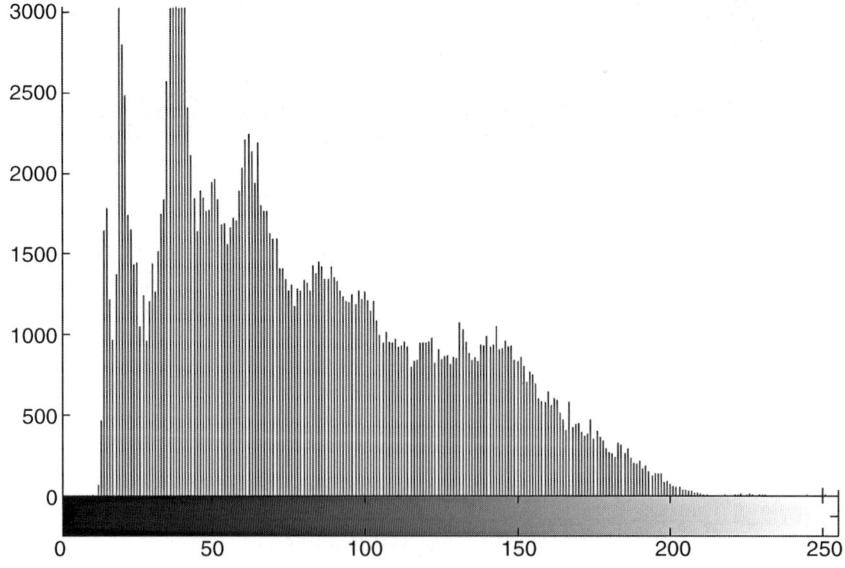


**Fig. 18.** Histogram of the image of Fig. 1.

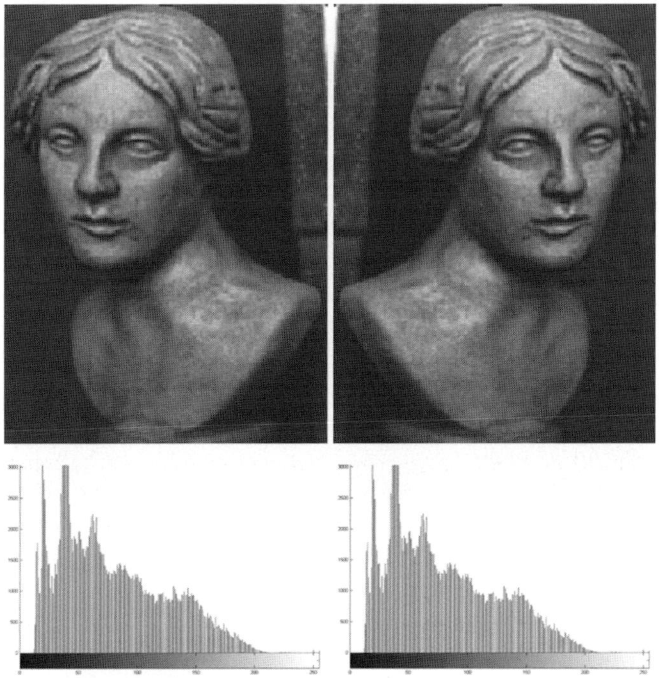

**Fig. 19.** Two different images (symmetric in this case) can have the same histogram.

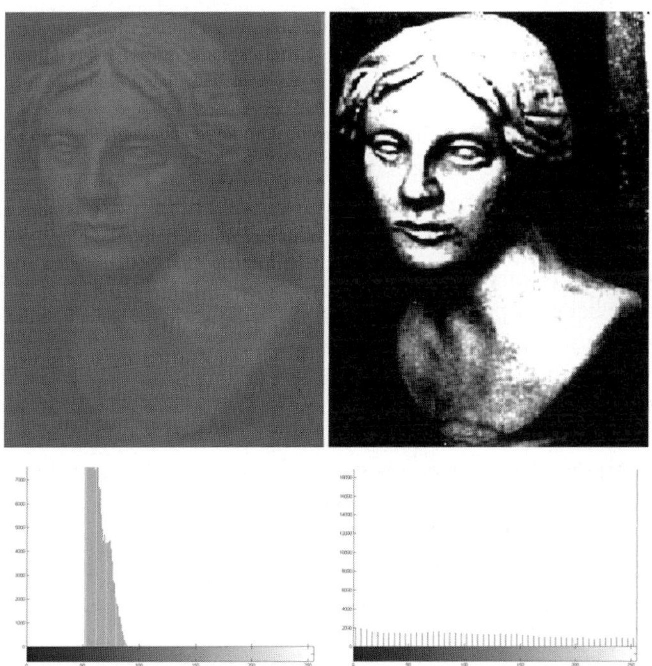

**Fig. 20.** Image with (left) low and (right) high contrast.

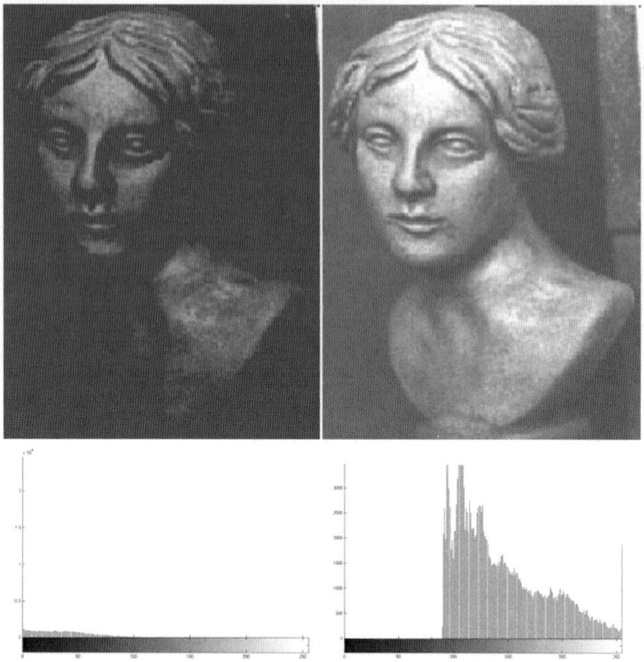

**Fig. 21.** Image with (left) low and (right) high brightness.

This is a "punctual operation" in the sense that it is performed on a pixel-by-pixel basis, without taking into account the "information" given by its neighbours.

*4.3.2. Contrast enhancement*

Working on histograms is very important for contrast enhancement. Many transformation laws ($S = T(r)$) have been proposed [13], however the simplest approach is to use the linear transformation shown in Fig. 22 where $T(r)$ is characterised by two parameters, $L_1$ and $L_2$ (the grey range is normalised from 0 to 1) only. When $L_1 = 0$ and $L_2 = 1$, the transformation is an identity (45° straight line); when $L_1 = L_2 = 0.5$, a bit-map is obtained (black and white pixels only). Other important laws are "logarithmic" and "exponential" transformation.

Another operation that enhances contrast is *histogram equalisation* (Fig. 22). Histogram equalisation aims at stretching the grey levels until they uniformly cover the entire intensity range. At the end of the operation, the histogram of the new image will be *flat*. This is true for a uniform intensity distribution. As we have discretised the intensity range, the equalised histogram will not be entirely flat and some grey levels will be lost. However, the values are far more uniformly distributed from black to white than in the original histogram and the contrast of the image is increased

Figure 23 shows the equalised histogram, and the relative transformed image. A comparison with Fig. 1 clearly shows that the contrast has been enhanced.

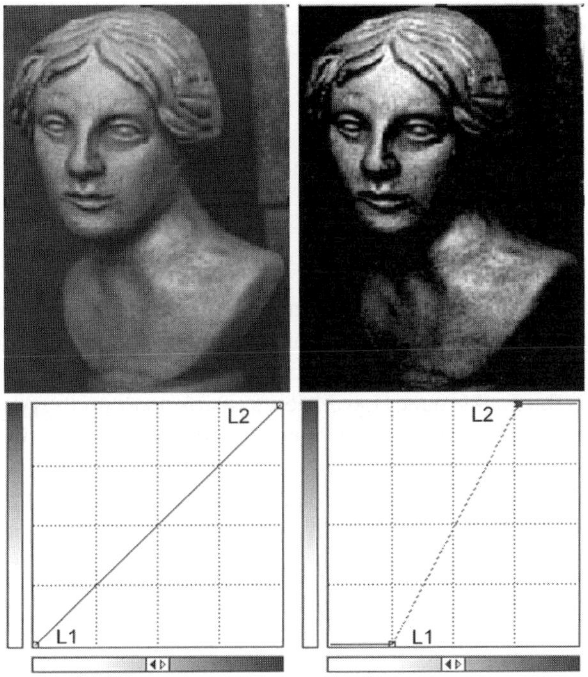


**Fig. 22.** Histogram equalisation, $T(r)$ as a linear function.

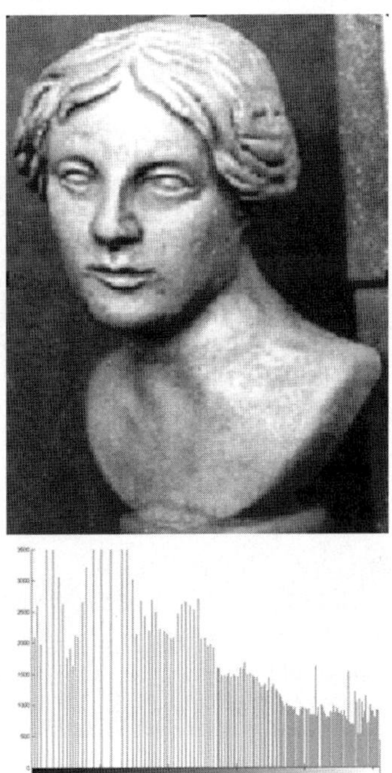

**Fig. 23.** Equalised histogram of Fig. 1.

$$T(r) = \begin{cases} 0 & 0 \le r \le L_1 \\ \dfrac{r - L_1}{L_2 - L_1} & L_1 \le r \le L_2 \\ 1 & L_2 \le r \le 1 \end{cases} \tag{10}$$

*4.3.3. Segmentation*

*Segmentation* is often used for a better rendering of an image with a bi-modal histogram [14]. This operation tends to detach parts of the image from the background for a better identification of them. An example is the bit-map of Fig. 17 (right-bottom). The "dark-grey" pixels, belonging to the background, have been transformed into "very-dark-grey" and the "white-grey" pixels into "very-white-grey" ones. Segmentation can also be performed for three-dimensional images. Having calculated the 3D distribution of the attenuation coefficient of the materials by computed tomography (see Section 8), one can set one material at zero density to make it completely transparent. Figure 24 shows an Etruscan bronze fibula, filled with the inner clay mould. By segmentation, applied to the histogram of the 3D image, it is possible to make the inner clay mould transparent, in order to obtain a better description of the inner bronze structure (Fig. 25) [15].

*4.3.4. Frame summing*

Digital images are often affected by random noise from many sources, such as the intrinsic fluctuation of photons (either from an X-ray beam or from the scintillator), the response of each pixel of the CCD, the multiplication of photoelectrons in intensified systems, and so on.

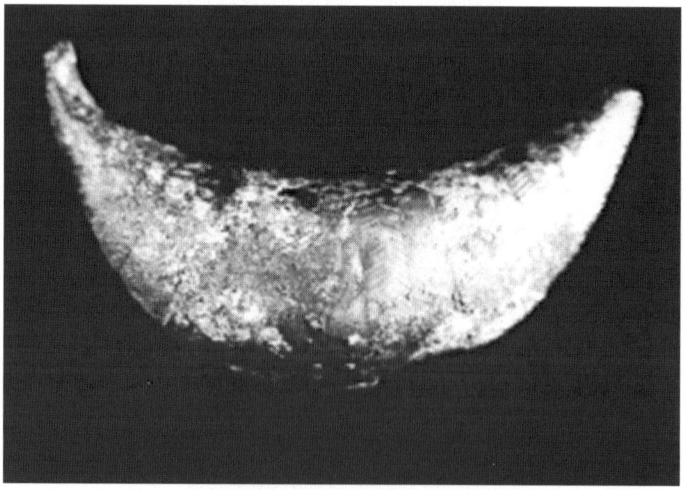

**Fig. 24.** Etruscan fibula (courtesy of Archaeological Museum of Bologna, Italy).

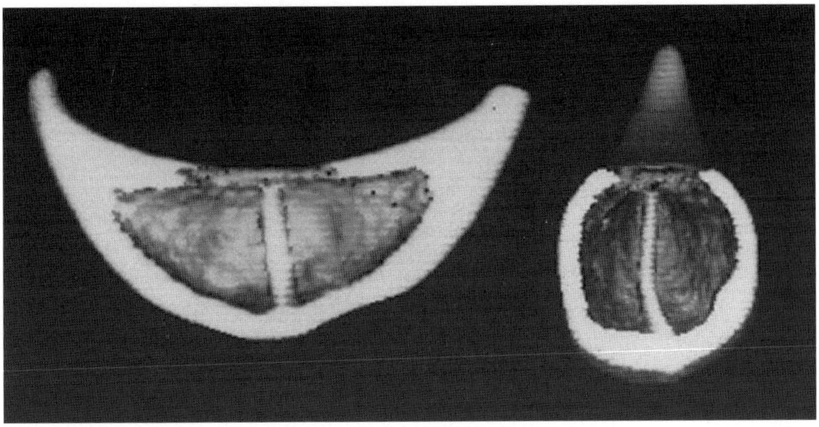

**Fig. 25.** CT of the Etruscan fibula (in collaboration with the Archaeological Museum of Bologna, Italy). The inner clay mould has been reduced to zero density (transparent).

This noise level determines the smallest intensity difference appreciable. Moreover, some pixels could be "blind".

One method that can be used to increase the signal-to-noise ratio (SNR) is to sum the digitised frames. This is equivalent to increasing the integration time thus decreasing the statistical fluctuation, and is only applicable for objects that have a stationary position within the frame provided the noise is uncorrelated between frames (truly random), and the SNR increases as $(N)^{1/2}$ where $N$ is the number of frames summed. Figure 26 shows the Fig. 1 with a "salt and pepper" noise added. Figure 27 shows the decreasing of noise after the application of frame summing ($N$ from 1 to 6).

**Warning!** It is not advisable for $N$ to be too high, as this could cause overflow and acquisition conditions could change if the time is too long.

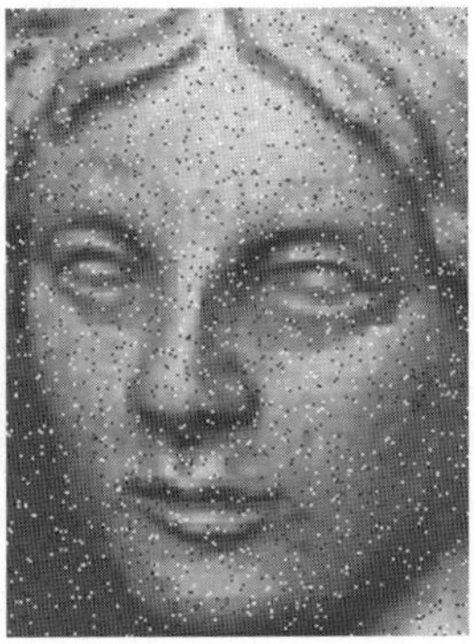

**Fig. 26.** Image with induced "salt and pepper" noise.

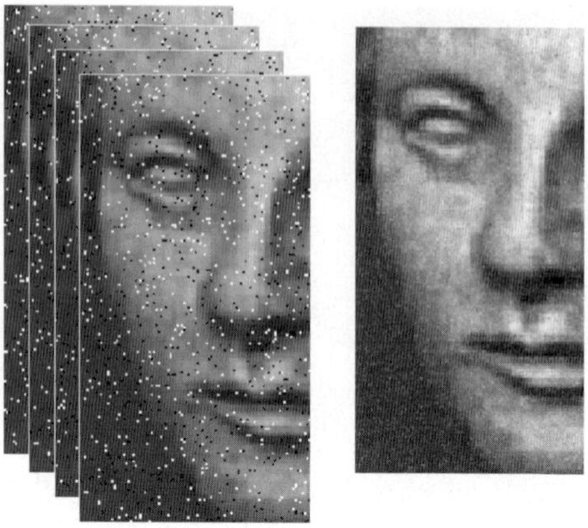

**Fig. 27.** Sum of different frames (up to 6).

*4.3.5. Pixel binning*

Detectors, most commonly linear arrays, may have a large number of pixels so that, sometimes, the image has a spatial resolution larger than desired. In such cases, it is appropriate to add the pixels of square or rectangular assemblies (usually $2\times2$ or $3\times3$) and to take the sum as a new value. Figure 28 shows images, binned in different ways. The elementary pixel increases in dimension but statistical fluctuation is considerably reduced.

## 4.4. Spatial filters

*4.4.1. Introduction*

A digital image can be "restored" just like an old painting or a noisy vinyl record. Many algorithms, known as "mathematical filters", have been developed for performing digital

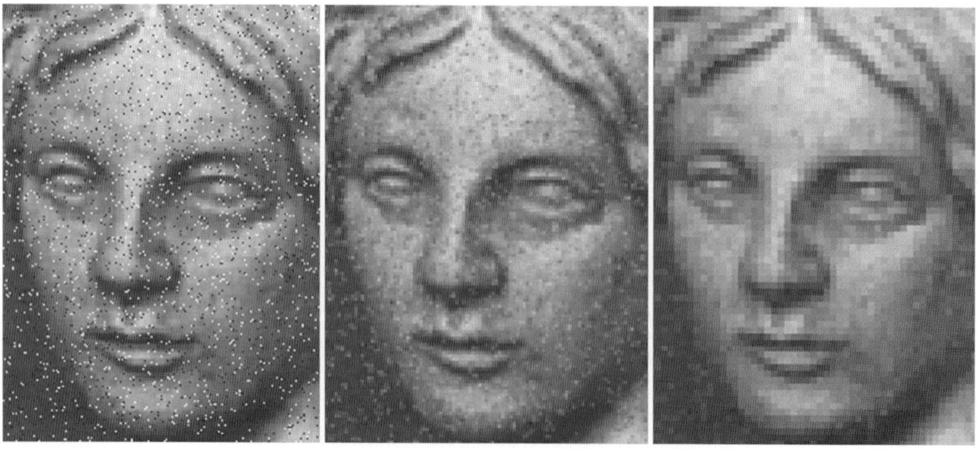

**Fig. 28.** The binning procedure reduces noise but increases pixel dimensions.

image restoration. Local filters and the Fourier Transform method will be discussed briefly here and readers wishing to know more about this subject should refer to specialised literature [16–18].

### *4.4.2. Image enhancement in the spatial domain*

Two important local filters are those connected to the *smoothing* and *edge enhancement* of digital images.

*Smoothing* is a process whereby noise is eliminated or decreased. One procedure entails the modification of the grey level of a pixel taking into account the grey levels of the neighbouring ones ("local treatment"). For instance, a white pixel in a dark image is likely to be a mistake. The white pixel can then be substituted by the mean grey value of the surrounding pixels, properly weighted. This technique is known as "linear spatial filtering". The larger the number of surrounding pixels, the smoother the transformed image will be. As the "smearing" of the noise decreases contrast, one can set a threshold and apply this transformation only if the difference between the original value and the transformed one exceeds this threshold. This procedure is named "medium filtering with threshold". Another approach, which is extremely useful in the case of "salt and pepper" type noise, is the so-called "median filter", consisting in taking the pixels of the neighbourhood being processed, ordering their grey values from smallest to largest, then taking the median value and using it to replace the original one (Fig. 29). Unlike the "medium filter", this filter maintains the boundaries but is not of a linear type.

*Enhancement* is a process whereby the difference between the zones of the image is enhanced. As smoothing is a sort of averaging (integral over a zone), enhancement will be obtained by applying the inverse operation, which is the derivative (gradient or laplacian operators). A threshold may also be imposed for these filters. For an optimal result, it is advisable to operate using several filters in sequence [19,20].

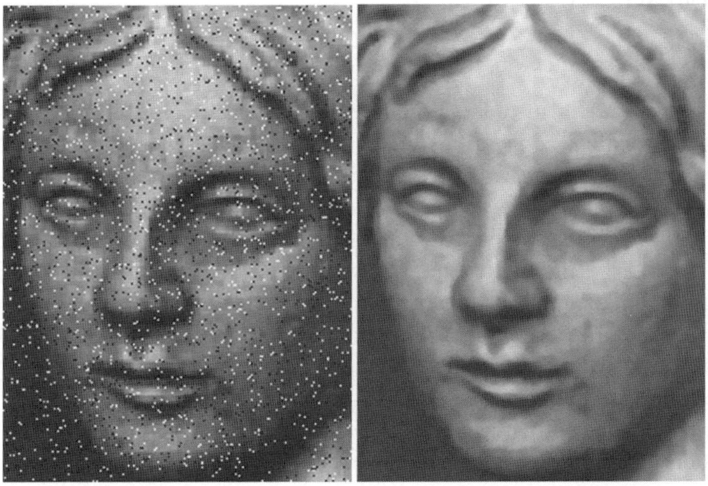

**Fig. 29.** Figure 26 "cleaned" using a median filter.

### *4.4.3. Fourier-Transform-based filtering*

For a better comprehension of this mathematical approach, let us start from the treatment of an acoustic signal, for instance a sound produced by an orchestra. Sound can be intended as a combination of many harmonics with different frequencies: the cello is characterised by low frequency harmonics, violins are characterised by higher frequency ones.

The whole of harmonics is called as the spectrum, which completely characterises the sound. If we use a proper electronic filter to decrease the high frequencies, we conversely increase the importance of the lower ones, that is the cellos. By decreasing the lower ones, we enhance the violins, and so on. The noise of an old vinyl record is also characterised by high frequency harmonics. So, if we develop in harmonics the sound of the record, then apply a filter to reduce the higher frequencies, we "clean" the sound of the noise. This mathematical procedure derives from the well-known development in Fourier series (see Appendix A).

One can proceed in much the same way with digital images. However, instead of using the development in Fourier series, and as a digital image is a discrete function, a more suitable mathematical approach, the FFT algorithm (Fast Fourier Transform), is used [21]. We can now modify the spectrum. If we decrease the high frequencies and let the low ones pass, we decrease the noise (equivalent to smoothing): conversely, if we decrease the low frequencies, we increase the importance of the rapid spatial variation, *i.e.* we enhance the boundaries. For images with unique characteristics, it is possible "to design" specific digital filters.

Unlike local filters, Fourier filtering treats images as a whole. These matters are dealt with in greater detail in Appendix A and, to an even greater extent, in the many books dedicated to them [18,22].

Mathematical software is commercially available to perform all these filtering operations.

## 5. DETECTORS FOR X- AND $\gamma$ RAYS

### 5.1. Families of detectors

It is possible to classify detectors into seven families.

1. *Gas-filled detectors* (for instance, argon at high pressure). These appliances were used in systems for medical applications. They have a very low efficiency but a very high dynamic range. Nowadays, they are used primarily in a number of industrial applications.
2. *CCD (Charge Coupled Device)-based detectors* are constructed of a semiconductor, usually silicon, in which the light produces pairs of electrons and vacancies. The CCD is like a pixel matrix; the higher the photon number, the higher is the charge collected in the single pixel. By measuring the charge collected in each pixel, and representing the measured value in binary form, one obtains a digital image. CCDs are also sensitive to X-ray photons that arrive directly on the silicon matrix. In such cases, the image is affected by undesired white zingers.
3. *Scintillation detectors*, consisting of a fluorescent material which emits light when exposed to X-radiation (*e.g.* CsI and $Gd_2O_2S$), are very widely used. The fluorescent material:
   (a) can be smeared directly (or indirectly through optical fibres) over a light detector (for instance, photodiode arrays or photomultiplier);

(b) can be smeared over a screen optically coupled to a CCD camera by a lens. A mirror, usually angled at 45°, makes it possible to keep the CCD camera out of the beam.

4. *Semiconductor detectors* (*e.g.* CdTe, CdZnTe, HgI, and Ge) allow direct photon counting with its energy, if required. Using this type of detector, it is possible to perform "gamma spectrometry". This type of equipment can be used for high energy X-ray imaging, which has a high dynamic range but a low spatial resolution (pixels no smaller than 0.5 mm).
5. *Image intensifiers* (I.I.) are based on rare earth screens from which the X-ray photons extract electrons, which, in turn, are accelerated by an electric field onto a fluorescent screen. A very bright image forms on the screen and is acquired by a CCD camera through a lens. Using I.I., it is possible to obtain digital images with low dose levels and therefore, they are often used in medical diagnostics. They have a low dynamic range and certain image distortions. One type of I.I. is the EBCCD (*Electron Bombarded* CCD) in which the extracted electrons are directly accelerated against a CCD without the lens coupling [23]. EBCCDs have smaller dimensions than standard I.I.s. Moreover, EBCCD can be fitted with a lens extending the range of the field of view (from small to large light source as a scintillating screen $30\times40$ cm$^2$).
6. *Flat panel* is a radiation detector of planar geometry which consists of a matrix of very small detectors (pixel with sides of 100 micron or less). They are made by amorphous selenium (Se-am) or with amorphous silicon (a-Si). In the Se-am version, the X-ray photons interact directly with the Se producing free charges, which are read by suitable electronics. In the a-Si version, a layer of scintillator (*e.g.* GOS or CsI) produces light which is read by a matrix of underlying sensors when bombarded by X-rays [24].
7. *CMOS* (*Complementary Metal Oxide Semiconductor*) is very similar to a flat panel. It is mainly composed of a matrix of microprocessors covered by a layer of scintillator (typically GOS or CsI). The light produced by X-ray interaction is transformed into electrical signals read by the underlying microprocessors. This type of detector needs a small amount of energy, which makes it suitable for transportable equipment, and are very fast. The pixel side is of the order of 90 microns and they can be assembled in "buttable" mode. This type of equipment could become the detectors of the future, at least for low energy photons.

Appendix C shows the characteristics of some flat panels, now (end 2004) on the market, and the features of a system based on a CCD camera coupled with a scintillating screen.

## 5.2. Geometry of the detection systems

Digital image acquisition systems can be listed in several ways. Below, reference will be made to their geometrical shapes: single detector, linear array of detectors, (planar) two-dimension detector.

### *5.2.1. Single detector (point geometry)*

Usually in this case, the detector receives the radiation through a narrow collimator (Fig. 30). This detection assembly is very useful in decreasing diffused radiation. Very good systems

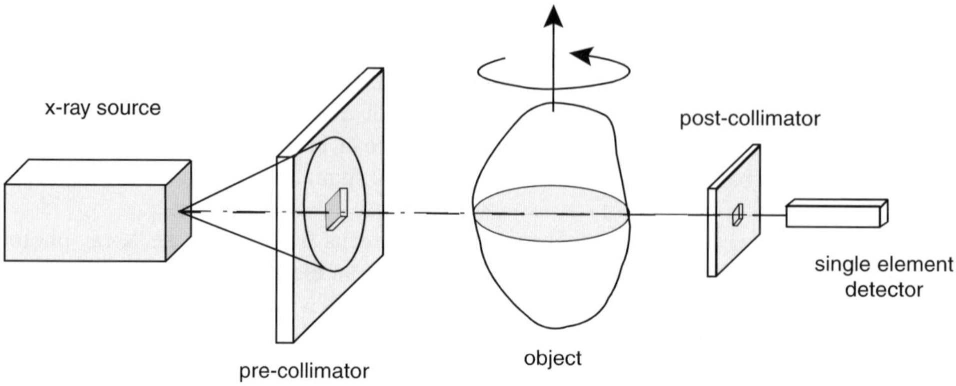


**Fig. 30.** "Pencil-beam" acquisition system with a single element detector.

are available on the market, mainly for medium and high-energy X-ray CT, with many single detectors each of which is well-collimated. The images are very sharp but acquisition times are rather long.

*5.2.2. Linear geometry (linear array)*

This assembly is also known as a "linear array detector"; it is composed of several single detectors (of the order of one thousand or more) positioned close to one another (Fig. 31) so that a "line" of object under investigation is obtained with a radiation shot. By moving the object linearly in front of the detector and "adding" the single lines, we obtain a matrix of pixels, that is a digital image as described in Section 4.2. The equipment used to check hand luggage in airports, has one or two linear array detectors (for looking from different angles). The source is usually collimated through a slit (fan beam). The collimation of the whole system is not as good as in the previous case but there is the advantage of a faster acquisition speed.

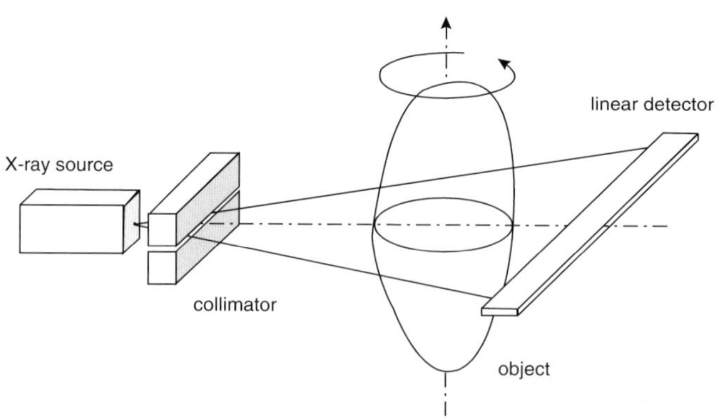


**Fig. 31.** "Fan beam" acquisition system with a collimated beam and a linear array of detectors.

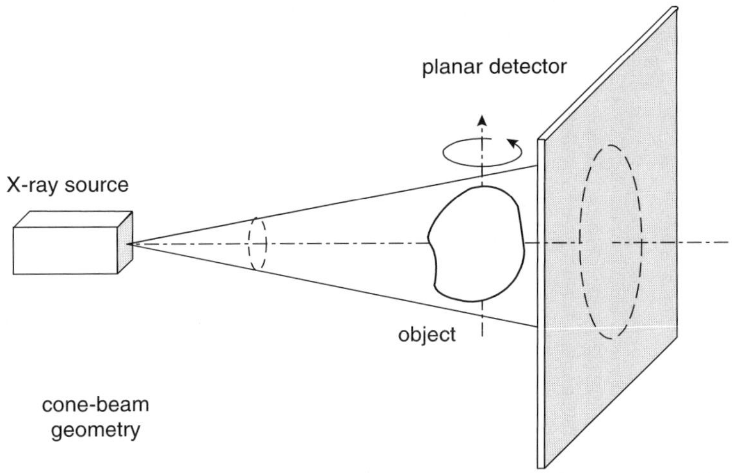


**Fig. 32.** "Cone beam" acquisition system with a broad beam and a planar detector.

By rotating the object in steps and making a radiation shot for each angular step, it is possible, after proper mathematical treatment, to reconstruct a "slice" of the body. A contemporary translation and rotation of the object gives so-called "spiral CT". Modern medical CT equipment is always of the "spiral" type [25].

Often, instead of having a single array of detectors, many detectors are packed together. In this case, we have multi-slice CT.

*5.2.3. Bidimensional geometry (planar detector)*

The image (shadow) is produced by a broad beam, named *cone beam*, over a planar detector, which can be a flat panel, a CMOS or a scintillating screen viewed by a CCD camera (Fig. 32). With this kind of system, a digital radiography (DR) is obtained with a single shot. By rotating the object and acquiring several DRs, after proper mathematical treatment (see Section 8.2), three-dimensional tomography of the object is obtained (*3D cone beam CT*).

## 5.3. The Modulation Transfer Function (MTF)

If we have a hi-fi radio, we can correctly reproduce either low or high frequencies (*e.g.* drums). However, if we have bad equipment, we will not be able to obtain quality sound reproduction, especially for high frequencies, as our radio "cuts out" part of the high frequencies. The same happens with a camera lens if the details of an image are too close. In order to quantify the quality of the lens, we can give the number of pairs of lines (succession of black and white) that our lens is able to separate. For this reason the spatial resolution of a lens is given in lp/mm (line-pairs per mm), which is a "spatial" frequency. The same approach is adopted for digital acquisition systems. Figure 33 shows a sequence of lead bars named

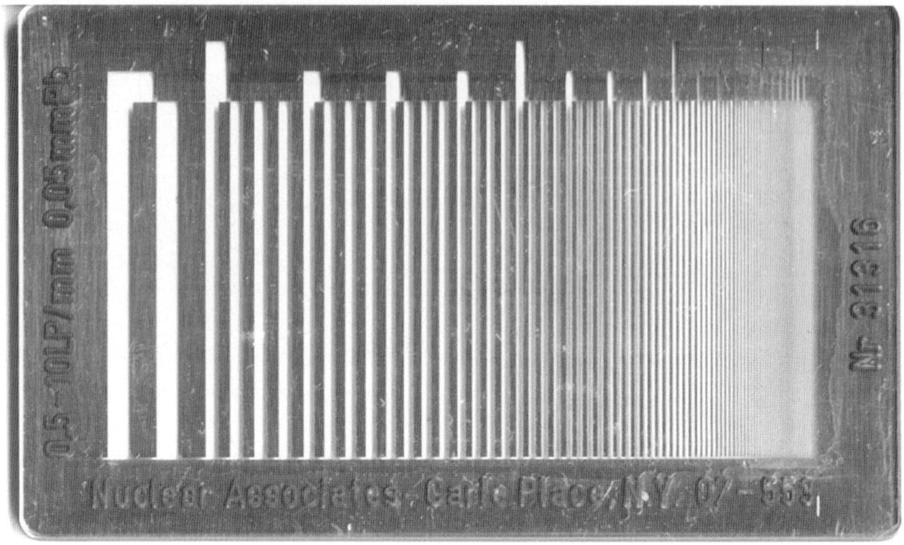


**Fig. 33.** Photograph of a line-pair gauge.

"line-pair gauge" with decreasing distances between one another. If we take an X-ray radiograph of this line-pair gauge (see Fig. 35) we will see that over a certain spatial frequency, our system is not able to separate the lead from the void. The Modulation Transfer Function (MTF) indicates the percentage of a modulated signal that our system will allow to pass. The higher the frequency, the lower the percentage will be. It is therefore possible, for each system, to create this function, which generally speaking, will decrease monotonically with spatial frequency, as in Fig. 34.

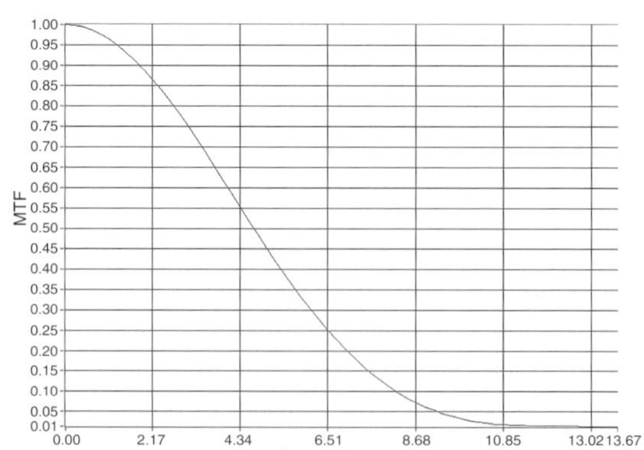


**Fig. 34.** Example of MTF.

By observing this figure, we notice that at the frequency 13.7 lp/mm, the MTF becomes zero (the "cut" frequency); moreover MTF is 0.5 (50% of signal passes) at 4.6 lp/mm and it is 0.05 (5%) at about 9 lp/mm. Therefore, when speaking of the MTF of a system, one must specify what the assumed percentage is. For instance, if a system has an MTF (5%) of 2 lp/mm, it means that it is possible to distinguish a detail of the dimension of 250 μm with a difference in grey of 5% from the background. If we have a complex system, like a chain for DR acquisition (X-ray source, scintillating screen, CCD camera, computer monitor and so on), the MTF of the system will be the product of the MTFs of each component.

**Warning!** It is useless to have one very good system component when another has poor qualities. If we consider a system with 2 components; the first of which has an MTF of 0.2 and the second an MTF of 0.8, then the system's MTF is 0.16. If we now consider a second system similar to the first, with components of MTF 0.5 and 0.5 respectively, the total MTF of this system is 0.25, that is better than the previous one.

Figure 35 shows the DR of the line-pair gauge, taken by a CMOS detector with a microfocus at 110 kV and 1 mA [26]. From this figure, it is possible to evaluate the MTF of the system.

Methods for calculating and measuring the MTF of a system are given in Appendix B.

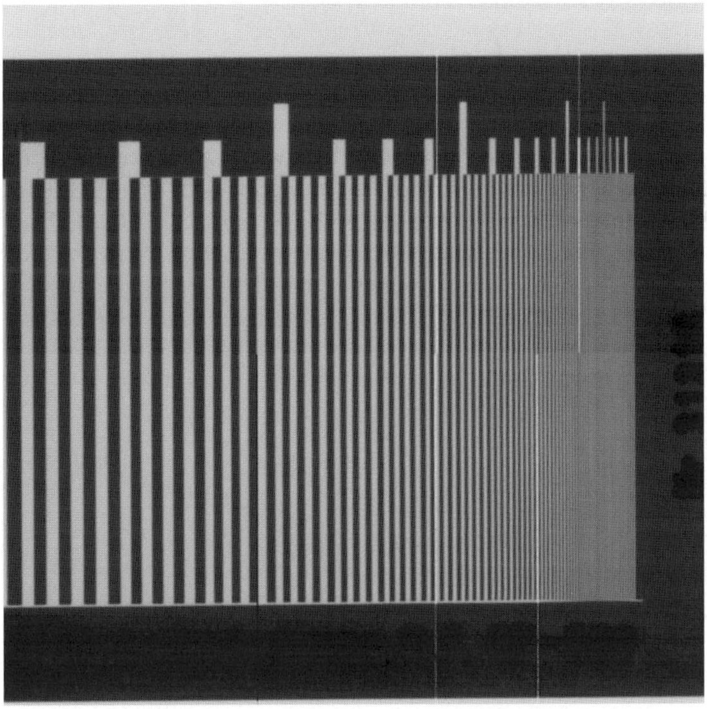

**Fig. 35.** X-ray radiography of line-pair gauge of Fig. 33.

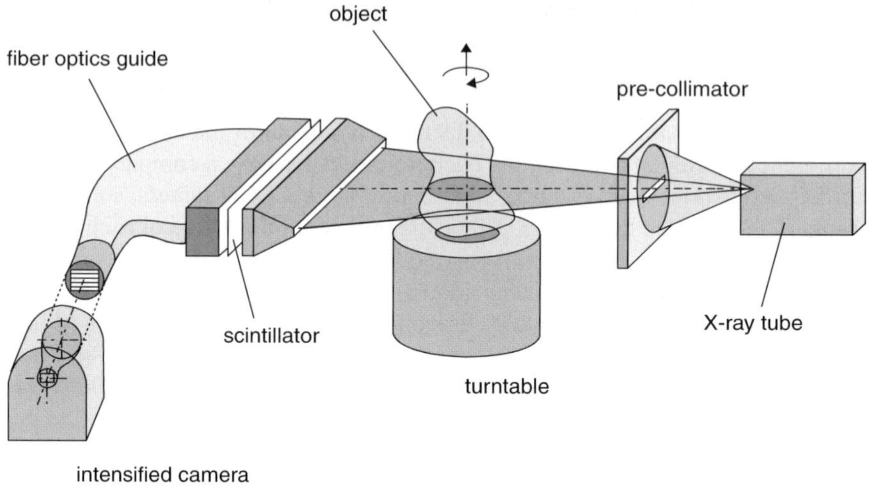


**Fig. 36.** Diagram of the linear array system.

# 6. EXPERIMENTAL ACQUISITION OF DIGITAL RADIOGRAPHS: SOME EXAMPLES

## 6.1. Acquisition by linear arrays

Detectors of linear array type are often used to obtain high resolution Digital Radiographs. A sketch of a new linear detector, developed by the University of Bologna, is shown in Fig. 36. It consists of a fibre optic (FO) fan that transports light over the photocathode of an EBCCD. The fan is made of seven ribbons as positioned in Fig. 37; this FO fan is a "geometry transducer", in the sense that it changes the geometry of the active area of the EBCCD from $1024 \times 512$ to $5607 \times 60$ pixels, thus obtaining a large (5600 pixels), multi-slice (60) detector.

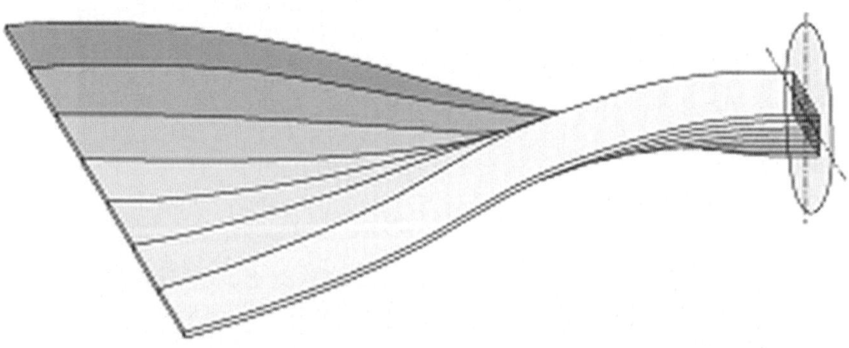

**Fig. 37.** Picture of the linear detector. An FO fan is coupled with the photocathode of an EBCCD.

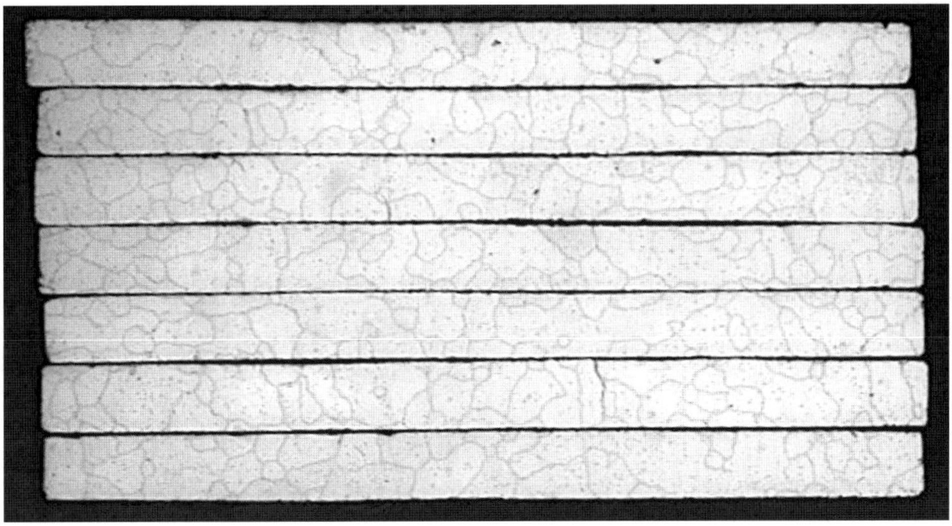

**Fig. 38.** The rear of an FO fan with patterns due to construction.

The image collected appears as a sequence of many slices that are rearranged to give one wide slice. The detector presents some "patterns" due to its construction feature (Fig. 38). The image acquired suffers this imperfection (Fig. 39), but it can be "cleaned" by subtracting the patterns (Fig. 40).

The DR is obtained as a sequence of many slices, as in the case of scanners. As the light converges over an EBCCD, the detector acts as an image intensifier so that the system can

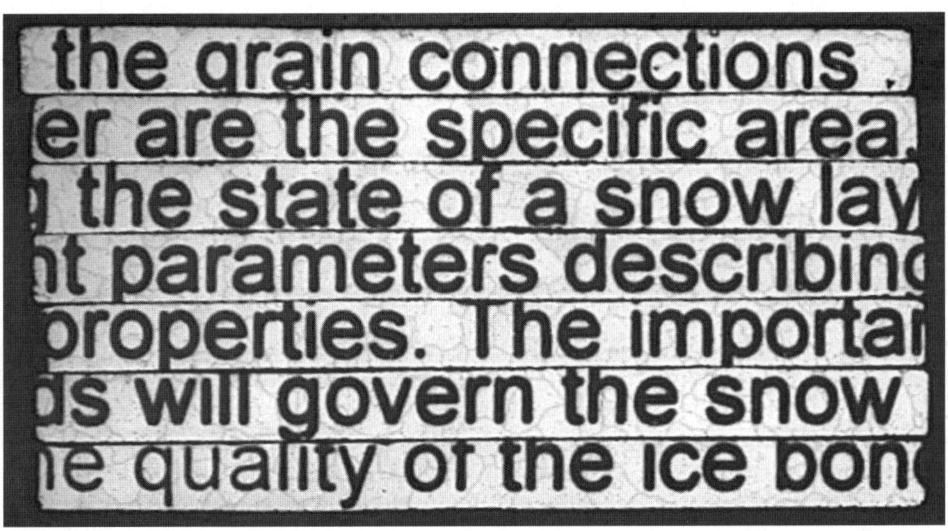


**Fig. 39.** The image affected by the pattern background.

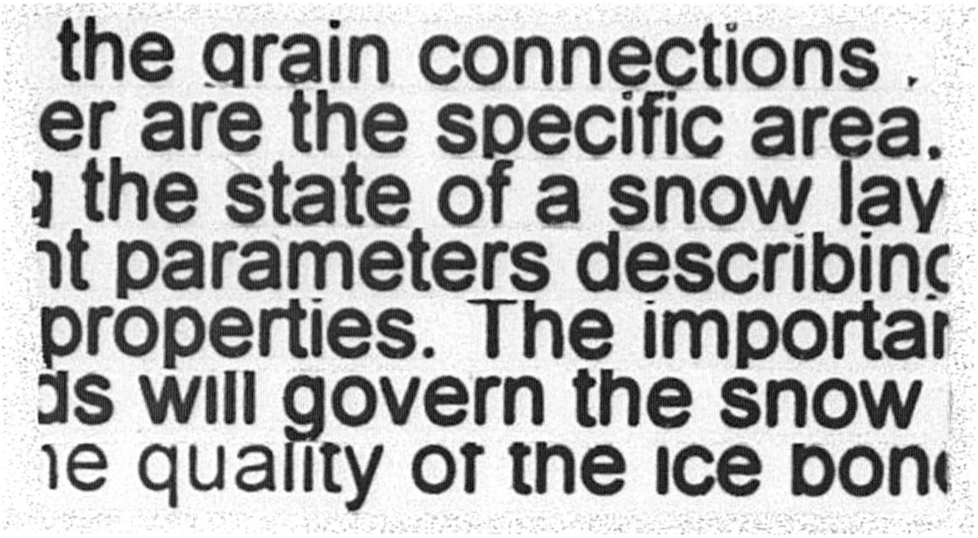


**Fig. 40.** The image of Fig. 39 "cleaned" of the pattern background.

perform DR with a far lower radiation dose than in standard films. If lower doses are essential for human beings, they are also advisable for paintings.

The aforementioned linear array was used as shown in Fig. 41 for the DR of a painting, a "test painting" with different pigments and cements prepared by the Opificio delle Pietre Dure in Florence. The high spatial definition of the detector allows the identification of the linen weft (see Fig. 42). Moreover, by making use of enhancement filters, it is possible to investigate either the painting or the frame with one shot alone, which is not possible with films that need different X-ray tube voltages. Another important feature of this detector is that the geometrical distortion, connected with film, is minimised with this geometry. By moving the detector and the X-ray tube synchronously with suitable equipment, it would be possible to easily "scan" a large painting and store all the high definition data on electronic media (*e.g.* DVD).

### 6.2. Acquisition using planar detectors

Of the many planar detectors suitable for acquiring digital images, the system that the Department of Physics of the University of Bologna and the Getty Conservation Institute have jointly developed is described below. It was designed for digital radiography (DR) and computed tomography (CT) to analyse objects of artistic interest. The system, designed to be used with a 450 kV X-ray tube, consists of an L-shaped aluminium box with a scintillator screen ($44 \times 44$ cm$^2$) of CsI(Tl), 1 mm thick, a mirror and a cooled CCD camera (Fig. 43). The image formed by the X-ray beam on the screen is viewed by the CCD camera ($2184 \times 1472$ pixels) *via* the mirror angled at 45°. The CCD camera is

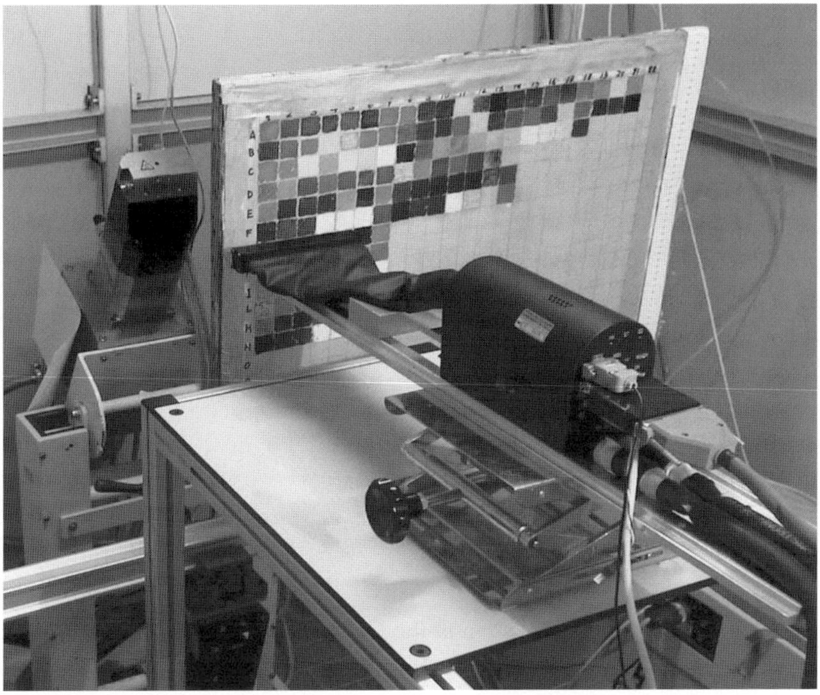

**Fig. 41.** The linear detector ready to perform a DR of a painting.

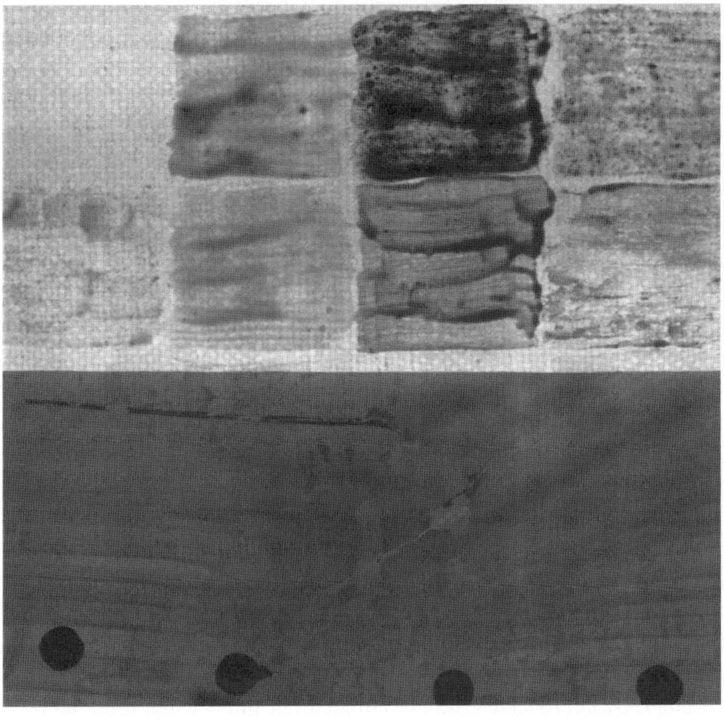

**Fig. 42.** Detail of the DR taken of the linen weft of the painting in Fig. 41.

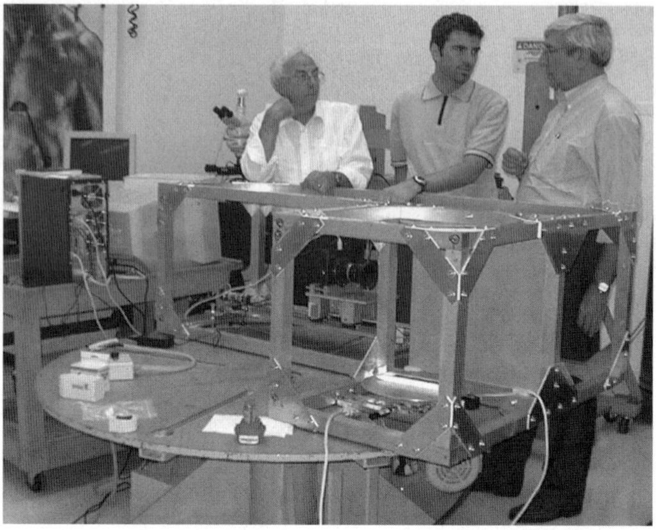

**Fig. 43.** The DR and CT system at the Getty Conservation Institute (GCI).

equipped with high aperture lenses. This feature of the system enables the detection of details smaller than 300 μm. Great effort was made to reduce the importance of the radiation diffused by the objects under investigation. This kind of radiation increases blurring and, conversely, decreases the contrast of images. The system has been tested on objects of different shapes and composition. Figure 44 shows a Roman bronze head in front of the scintillator and the Fig. 45 shows the acquired image. Computer tomographs, acquired by this system, are shown in Section 9.3.2. More details of this system are given in Appendix C.

## 6.3. The advantages and disadvantages of digitising

As mentioned above, a digital image is equivalent to a matrix of numbers and, therefore it can be saved (in CD or DVD) for a long time without degrading (in theory!), unlike the radiographic film. A digital image can be transmitted via the Internet and can be processed to reduce noise or increase the contrast as explained in Section 4 and in Appendix A. On the other hand, a digital image usually cannot achieve the spatial resolution of an analogue image. However, with modern CCD cameras, it is possible to get submicrometric resolution that cannot be achieved using film.

**Warning!** Since the 1950s, great changes have taken place in electronic storage with the introduction of magnetic tapes, optical disks, floppies, CDs, DVDs, which has led to great suffering and expenditure when one needs to recover something old (from image storing point of view) or when translating data from an old format to a new one. It is therefore preferable, when possible, to save data also in an analogue format that can be treated easily

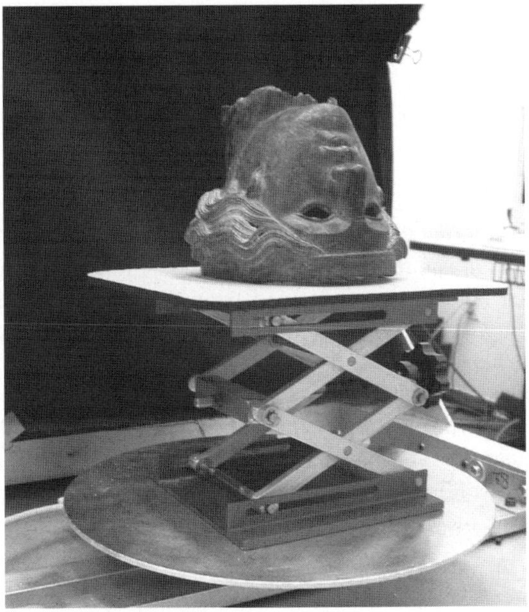

**Fig. 44.** Photo of the head of an ancient Roman statue in front of the planar detector at the GCI.

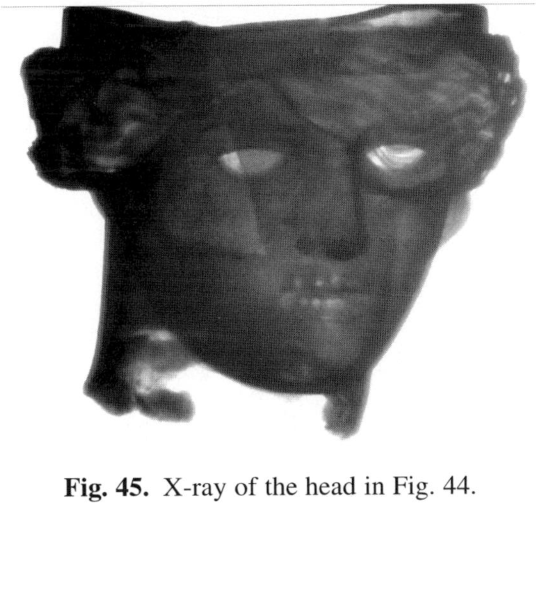

**Fig. 45.** X-ray of the head in Fig. 44.

in future or updated from old to new format should this still be feasible. In the future, this conversion could be dramatic!

## 7. DIGITAL IMAGING FOR NEUTRON RADIATION

### 7.1. General considerations

Unlike photons, which interact primarily with the electrons of atom shells, neutrons interact with atomic nuclei and as a result, a different kind of reaction is to be expected. When interacting with matter, neutrons are removed from a beam (by absorption or scattering) by light elements, like hydrogen, deuterium and carbon, or by particular isotopes with high capture cross sections, like $^{10}$B, $^{6}$Li and $^{155}$Gd. Figure 46 shows the linear absorption coefficient of some isotopes for neutrons and X-rays [27]. Using neutron radiography, it is possible to see a plastic film on a block of lead, which would be absolutely impossible using X-rays. From this point of view, neutron radiography may be considered as "complementary" to conventional radiography. As neutrons have no charge, their detection is based on the indirect ionisation they produce.

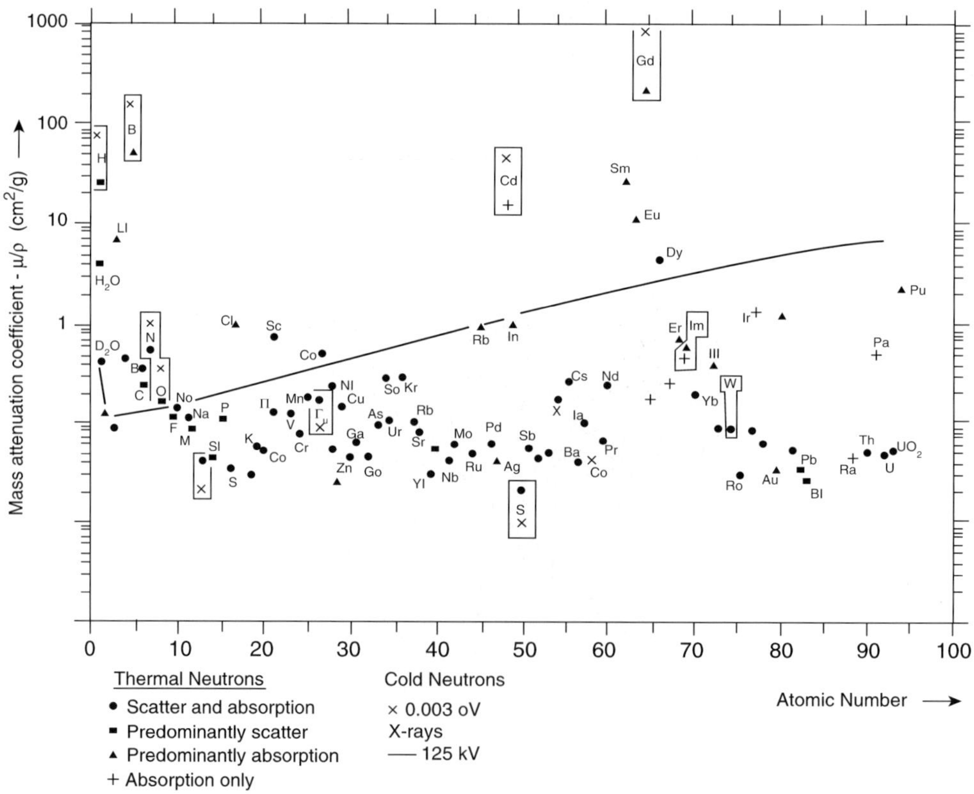


**Fig. 46.** Neutron and X-ray mass attenuation coefficients for neutrons and X-rays [27].

The most important reactions used, for imaging purposes, are:

(a) n + $^6$Li → $^3$H + α, with kinetic energy of reaction products of about 4.79 MeV;
(b) n + $^{10}$B → $^7$Li + α + γ with kinetic energy of reaction products of about 2.79 MeV;
(c) interaction [neutron → proton] with detection of proton;
(d) n + $^3$He → $^3$H + p, with kinetic energy of reaction products of about 0.764 MeV;
(e) interaction [neutron + fissile nucleus], with kinetic energy of fission products of about 200 MeV.

Reactions (a) and (b) are used for the detection of so-called "thermal" (with $E_n < 0.4$ eV) and "epithermal" neutrons (0.4 eV < $E_n$ < 100 keV); reaction (c) is for fast neutrons ($E_n > 0.5$ MeV), and reactions (d) and (e) are for both thermal and fast neutrons.

### 7.2. Planar detectors for neutrons

For digital neutron imaging, the most common detectors are planar models (see Section 5.2.3). In particular, for thermal neutrons, scintillating screens smeared with a mixture of $^6$Li and ZnS are used [28]. The neutron is absorbed by $^6$Li, thus producing an α particle and a triton which dissipate their kinetic energy in the ZnS, creating a light flash (see reaction (a)). The light can be recorded by a CCD camera [29]. For neutron radiography, $^{10}$B loaded film can also be used [30].

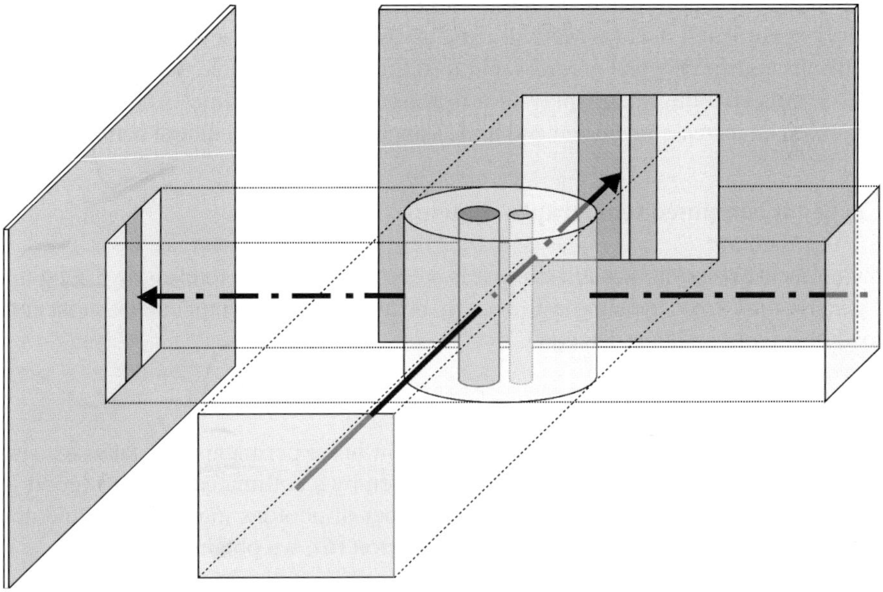

**Fig. 47.** Two projections, taken at right angles, show how some details are not visible in one of them. In order to overcome this drawback, computed tomography is based on a set of very many projections and enables the correct reconstruction of the complete section or volume.

As mentioned previously, the advantage of neutron imaging is that it is able to detect organic material traces in metallic objects such as corrosion or inner clay mould or internal wooden parts.

## 8. COMPUTED TOMOGRAPHY USING X-RAYS AND GAMMA PHOTONS

### 8.1. General considerations

Let us consider a cylindrical vessel containing two smaller cylinders. Two radiographs (projections) of this object are shown in Fig. 47; it goes without saying that what we are able to see of the content of the vessel depends on the projection angle. For instance, in the projection on the left, the smaller cylinder is covered by the larger one. If, instead of having projections on one plane, we virtually cut the container and consider its "section", known as a "slice", we can see both inner cylinders. This operation of "cutting" is named "tomography", from the Greek meaning "to cut".

The problem of how to obtain a section of an object using an infinite set of rays passing through it, was solved theoretically in 1917 by the Austrian mathematician Radon. However it was not until the 1960s that two scientists, the physiologist Geoffrey Hounsfield and the physicist Allan Cormack (separately) succeeded in obtaining the section of an object experimentally.[5]

The first equipment consisted of a gamma beam (a radioisotopic source, inside a collimator) impinging on a collimated detector. Hounsfield took several days to collect all the necessary experimental data (as we will explain elsewhere), and several more to process it. Today, thanks to technological progress, ultrafast CT systems are able to acquire slices in less than 0.2 s, thus enabling the display of a beating heart in real time [31]. In recent years, there has been an increase in the demand for CT applications in the cultural heritage sector too.

### 8.2. Types of computed tomography systems

Following their evolution through time, it is a common practice to classify CT equipment into "generations" from the simplest (a single beam and one detector) to the most complex (a broad beam and a planar detector).

#### *8.2.1. First generation CT system*

Following the procedure adopted by Hounsfield in his experiments, we take, for simplicity, a collimated monoenergetic gamma source seen by a collimated detector (good geometry conditions) as in Fig. 48. Let $N_0$ be the number of photons impinging on the detector having run the chord $d_1$ of the object. From equation (6), we obtain:

$$N(d_1) = N_0 \exp(-\mu d_1).$$

[5] For the development of Computed Aided Tomography, G.N. Hounsfield and Allan M. Cormack received the Nobel Price for Medicine in 1979.

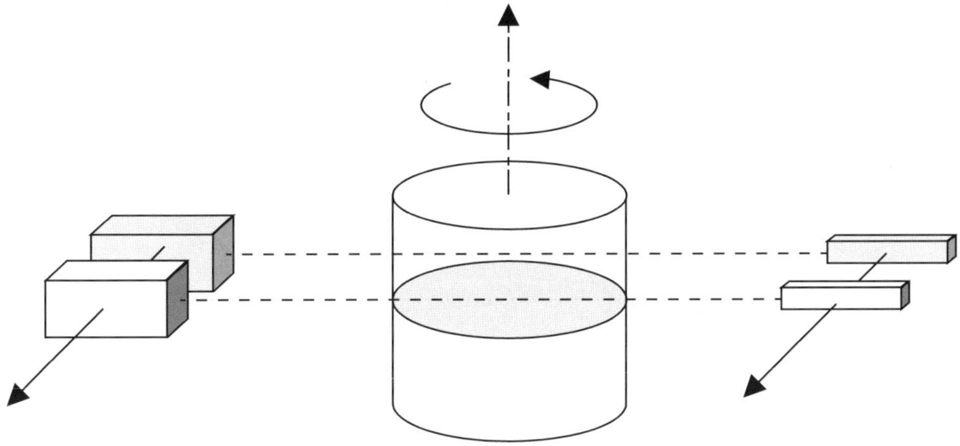


**Fig. 48.** First generation tomography system: a single detector scanning the object at each angle (translation and rotation).

From which one gets:

$$\ln(N_0/N(d_1)) = \mu d_1. \tag{11}$$

Taking into account that, in general, $\mu$ is a function of space ($\mu = \mu(x,y)$), equation (11) may be written as:

$$\ln(N_0/N(d_1)) = \int \mu(x, y_1)\mathrm{d}x. \tag{12}$$

By moving the "source-detector" system in relation to the object (or *vice versa*), the beam will cross the object through another chord, $d_2$, thus obtaining another experimental value:

$$\ln(N_0/N(d_2)) = \int \mu(x, y_2)\mathrm{d}x. \tag{13}$$

By repeating the same operation for many chords and rotating the object (or the "source–detector" system) by a small angular step ($\Delta\phi$), then repeating the measurements from the beginning several times, one obtains a "net of beam-rays" covering the whole slice under investigation. Using a proper mathematical procedure [32], it is possible to reconstruct the function $\mu(x,y)$, that is the absorption linear coefficient of the body under examination.

As the number of measurements cannot be infinite, the function $\mu(x,y)$ is not continuous but rather "pixelised", like a digital image. The smaller the linear step ($\Delta d$) and angular step ($\Delta\phi$), the better the $\mu(x,y)$ reproduction will be. As the beams are parallel, it is sufficient to rotate the object through 180°. This procedure is known as "first generation CT".

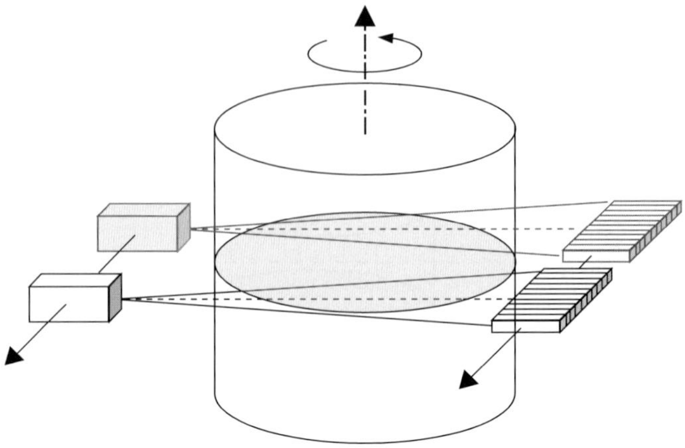


**Fig. 49.** Second generation tomography system: a linear array of detectors scanning the object at each angle through 360°, but with a lower shift number (translation and rotation).

***"Thumb rule"*** The number of angular steps should not be less than the number of linear steps.[6]

*8.2.2. Second generation CT system*

Although first generation systems are the most correct with regard to image reconstruction, they require lengthy timeframes. In order to reduce acquisition times, instead of a single detector, an array of *N* detectors is used (Fig. 49) (see Ref. [33]), which is equivalent to performing *N* measurements at the same time. With this type of detector, the translation step number decreases (but the number of rotation steps does not). This type of system is known as "second generation CT" equipment.

As, unlike first generation systems, the beams are not parallel, rotation must be performed through 360°. When a "synchrotron light" is used as a radiation source, the rays are parallel and consequentially, rotation is through 180°.

The equipment used for the CT of large objects (for instance, rockets), are often second generation models [34,35].

*8.2.3. Third generation CT system*

If the linear array is wide enough for the whole object to be projected over it (see Fig. 50), rotation alone is needed. If rotation and translation – in direction perpendicular to the rotation plane – are performed simultaneously, we can produce "spiral CT". This type of CT is suitable for cylindrical objects such as a column, rock core or human body.

[6] It can be demonstrated that, for having good results for the reproduction of the outer part of the object under examination, the optimal number of the angular steps should be about $\pi/2$ times the number of the linear steps.

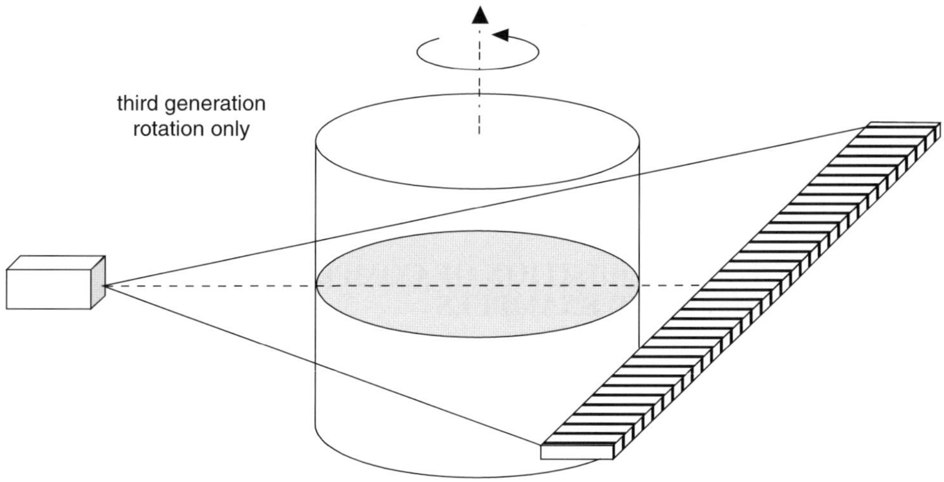


**Fig. 50.** Third generation tomography system: a wide linear array of detectors collects the projection of the complete section of the object at each angle through 360°. It is no longer necessary to move the detector (rotation only).

### *8.2.4. Medical CT*

Almost all new CT systems for medical diagnostics are of the spiral type. In this type of systems, the detectors are located on a circumference that surrounds the cavity in which the human body moves. The X-ray tube rotates continuously irradiating part of the detector ring. In modern CT, there are several rings of detectors, named "multi-slice system" (*e.g.* see Ref. [36]).

**Warning!** Medical CT is suitable for use in the cultural heritage field when the object examined has similar characteristics to those of a human body, such as Egyptian mummies, wooden statues, etc. it goes without saying that it is not possible to acquire tomographs of metal objects (such as bronze heads) due to the low penetration of the X-rays of these tubes (maximum voltage of the order of 160 kV).

### *8.2.5. "Cone beam" tomography*

To increase the acquisition speed, the source beam – in the shape of a cone – totally irradiates the object that rotates in front of a planar detector (Fig. 32). If the detector is smaller than the projection, a macro-slice of the object only is acquired. In order to obtain a CT of the entire object, it is necessary to move the object vertically and acquire several macro-slices that are later "joined" together using a dedicated software programme. This approach is called the "cone beam tomography" and it is often used to inspect pieces of archaeological interest. The disadvantage of cone beam tomography is the high percentage of diffused radiation that impinges on the detector, as no post-collimator is present. Like planar detectors, they can use flat panels, a scintillating screen seen by a CCD camera or CMOS covered by a suitable scintillator. The FDK algorithm is usually adopted for image reconstruction [37]. (FDK is an acronym of the intials of Feldkamp, Davis and Kress, the authors of [37].)

**Warning!** For this approximation to be valid, the aperture of the cone angle must be not greater than 10°.

## 9. EXPERIMENTAL ACQUISITION OF COMPUTED TOMOGRAPHS: SOME EXAMPLES

### 9.1. Foreword

Until a few years ago, computed tomography was a diagnostic technique applied mainly to human beings. With the development of different types of detectors and the lowering of costs, this technique is now also more widely used in the cultural heritage field. However unlike the human body, the dimensions of objects of cultural interest cover a wide range: from a prehistoric tooth, just a few millimetres long, to large globes with diameters of over 2000 mm. It is therefore necessary to develop different kinds of CT systems, each one specialised in a particular type of object. Descriptions are given below on the different kinds of equipment developed, considering cultural heritage requirements only.

### 9.2. Microtomography

#### *9.2.1. Microtomography in cone beam geometry*

If the object under investigation is small (few mm) and if a good spatial resolution is required (of the order of few microns), then a micro-tomographic system is used (μ-XCT). In this type of system, phosphor is smeared over FO tapers, or FO ribbons or directly over the CCD. A microfocus or nanofocus is used as an X-ray source (see Section 2.3.1). When available, synchrotron light constitutes a very efficient source (see Section 2.3.3). A μ-XCT system is shown in Fig. 51 [38]. It has a field of view of $30 \times 15$ mm$^2$. Figures 52 and 53 show the CT of an ancient Roman tooth and a fossilised jaw acquired using this type of CT system [39]. A fine focus tube was used as the X-ray source and it has a focal spot ranging from 5 to 100 μm (depending on the power), maximum voltage 200 kV and maximum current 2 mA.

#### *9.2.2. Microtomography with a linear detector*

The detector described in Section 6.1 was also used in CT configuration (see Fig. 38). Using a synchrotron ELECTRA (SYRMEP beam-line) as an X-ray source, it was possible to obtain a multi-slice CT of a human femur with a spatial resolution comparable to that of a small bone fragment (see Fig. 54) and this slice constitutes one of the largest objects tomographed with this kind of spatial resolution [40].

If one requires very high spatial resolution, it is possible to use a single crystal of scintillator viewed by a microscope equipped with a CCD camera.

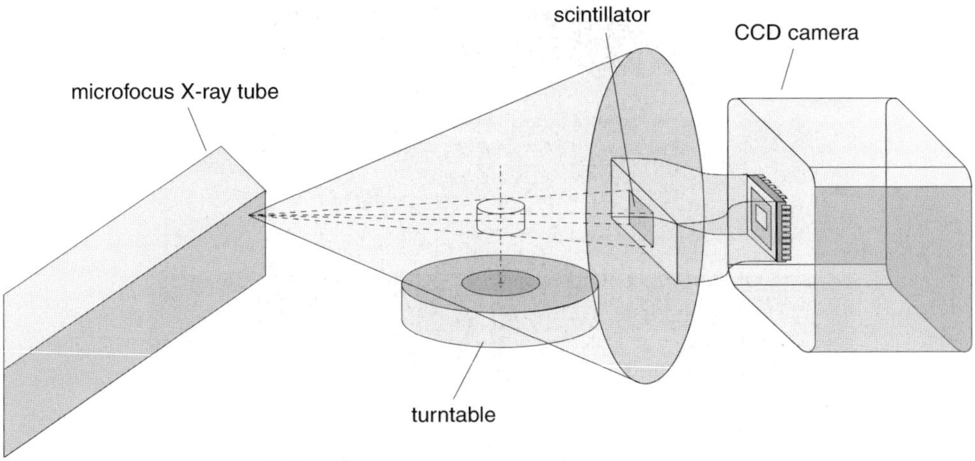


**Fig. 51.** Diagram of an experimental microtomography system. A cooled CCD camera is optically coupled with a scintillating material layer by means of a fibre optic taper in a cone-beam geometry.

### 9.3. Medium-size CT systems

Medium-size CT systems have screens ranging from $30 \times 30$ cm$^2$ to $40 \times 40$ cm$^2$, using either a flat panel (see Section 5.1) or a home-made system, the features of which can be adapted to suit user needs. Two home-made systems will be described here.

#### *9.3.1. CT system with EBCCD*

This system has a GOS scintillating screen $30 \times 30$ cm$^2$; the image produced on the screen by the X-ray beam is viewed by a $1024 \times 512$ pixel EBCCD camera. As this camera is

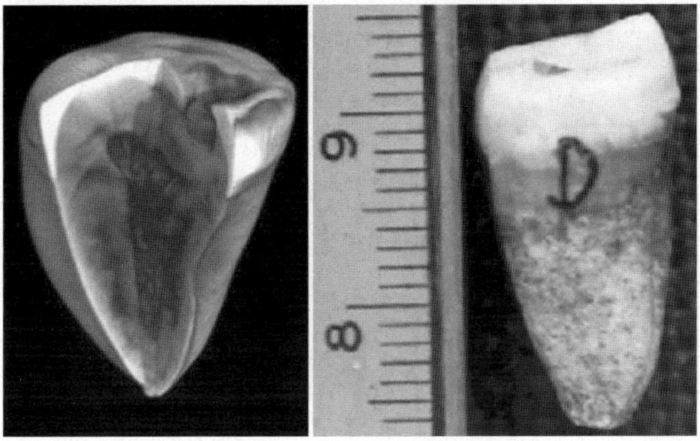

**Fig. 52.** Micro-CT of a Roman tooth.

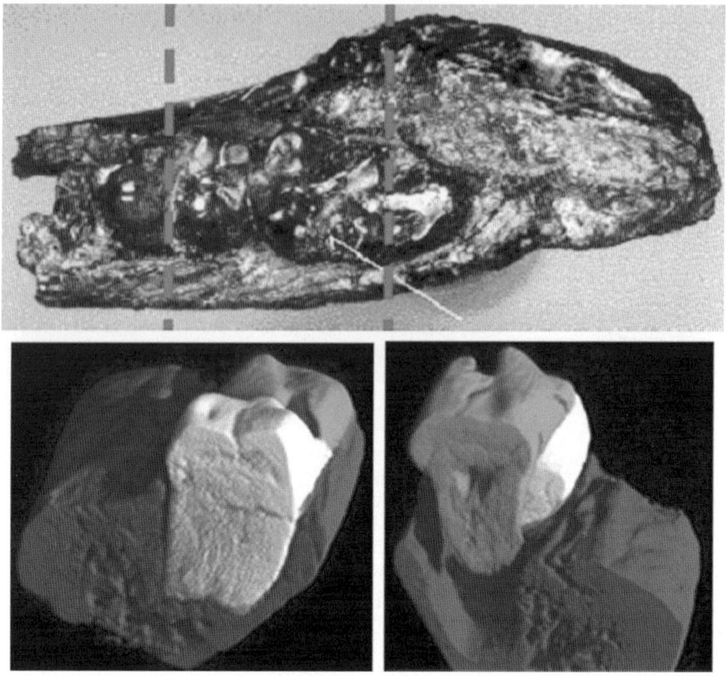

**Fig. 53.** Photo and micro-CT of a fossilised jaw.

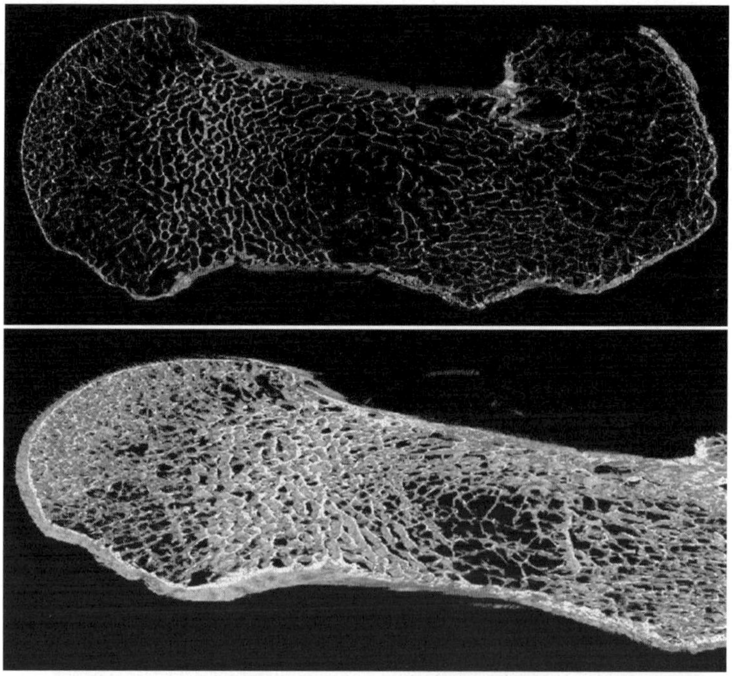

**Fig. 54.** CT of a human femur with a multi-slice detector (Fig. 38) and synchrotron light source.

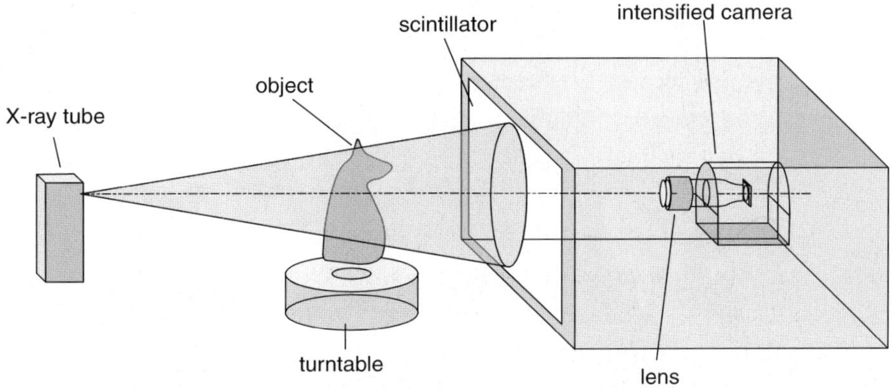


**Fig. 55.** An intensified camera collects the image produced by the X-rays on a scintillating material screen $30 \times 30$ cm$^2$, that represents the radiographic projection of the object.

intensified, the images need very low radiation intensity. It is therefore possible to acquire objects very quickly or acquire thick objects (see Fig. 55). This CT system has been used for the inspection of a mummified Egyptian cat (see Fig. 56) (supplied by the Archaeological Museum of Bologna). Using suitable software, it is possible to "remove" the cat's skeleton from its coffin (Fig. 57). This system operates well up to 300 kV.

*9.3.2. Medium-high energy*

This system is briefly described in Section 6.2. It has been tested by taking CT of objects of different shapes and compositions. Figures 58 to 60 show CT of a small bronze elephant, a wooden horse with an iron core and an ancient bronze head dating to Roman times. As the top part of the head is missing, the reconstructed image can also be visually

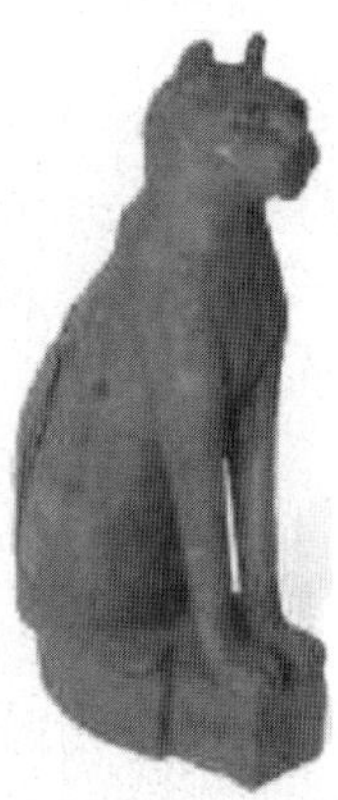

**Fig. 56.** Photo of an Egyptian coffin with a mummified cat (in collaboration with the Archaeological Museum of Bologna).

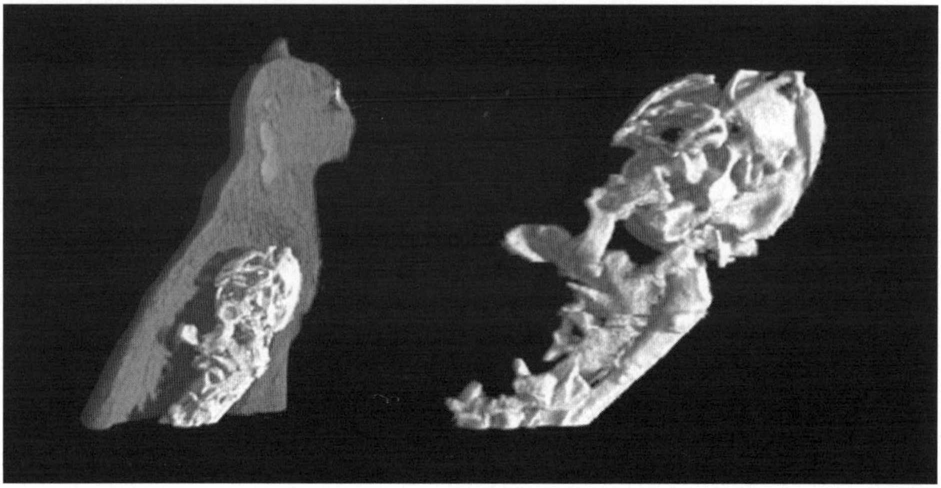

**Fig. 57.** CT of the coffin shown in Fig. 56 containing the skeleton of a cat.

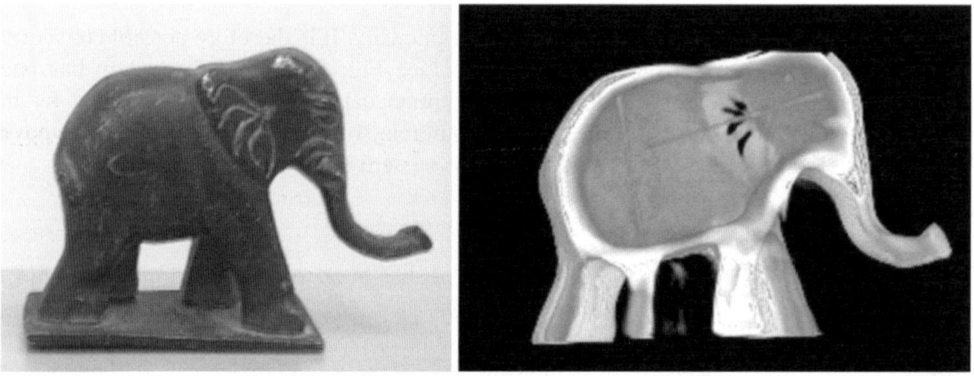

**Fig. 58.** CT of a small bronze elephant.

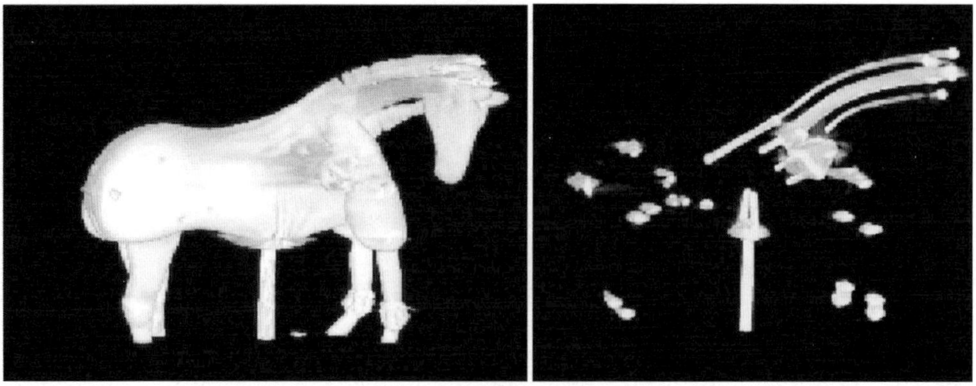

**Fig. 59.** CT of a wooden horse with iron core.

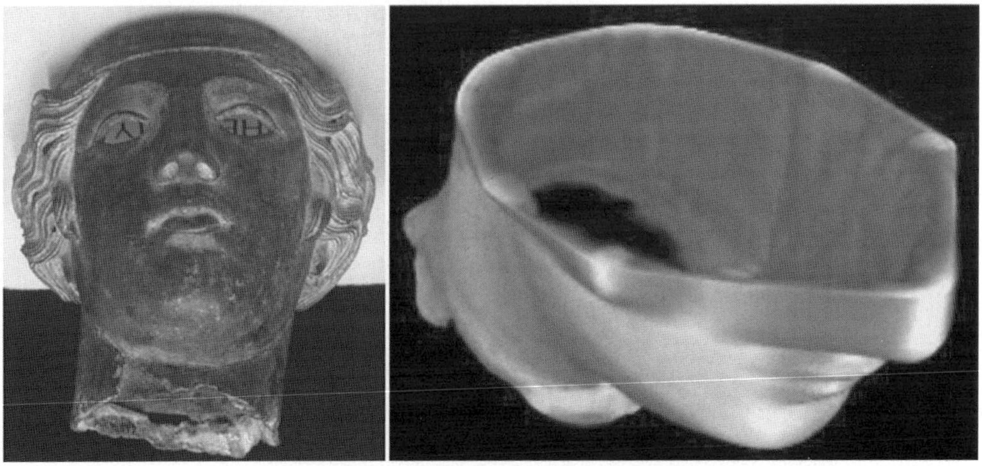

**Fig. 60.** A head of an ancient Roman bronze statue (see Fig. 44) and the 3D reconstruction thereof.

compared with the actual internal structure. This system can operate well up to 450 kV. Rather higher energy can be obtained by LINACs. With these X-ray sources metal objects can be inspected (for instance, see Refs. [11,40]).

**Warning!** Before irradiating a bronze statue (mainly for CT investigation), it is a good practice to keep part of the inner clay mould for conducting, the age measurement by thermo-luminescence technique. The irradiation, later on if desired by X-rays or $\gamma$ rays would perturb the measurement, and the statue would appear older.

### 9.4. Computed tomography of a large ancient globe

In Palazzo Vecchio, at Florence, there is a large globe (2200 mm in diameter) created by a Dominican monk, Egnazio Danti, around 1567 (see Fig. 61). The Municipality of Florence, in collaboration with "Opificio delle Pietre Dure" in Florence,[7] decided to set up an important diagnostic campaign for this wonderful masterpiece and repair, as much as possible, the injuries of time. Besides the cleaning of the surface, which had become brown, the project

[7] The diagnostic campagin was decided by the Municipality of Florence and by the "Opificio delle Pietre Dure" of Florence under the surveillance of "Sopraintendenza per i Beni Architettonici e il Paesaggio" and for the "Patrimonio Storico Artistico e Demoantropologico" of the Provinces of Florence, Pistoia and Prato. The diagnostic campaign was carried out by the National Institute of Applied Optics (INOA) of Florence, the Institute of Science and Information Technology (ISTI-CNR) of Pisa, the Department of Physics of the University of Bologna, the Department of Chemistry of the University of Perugia, and the Systems Measurements Services (S.M.S.) at Sutri.

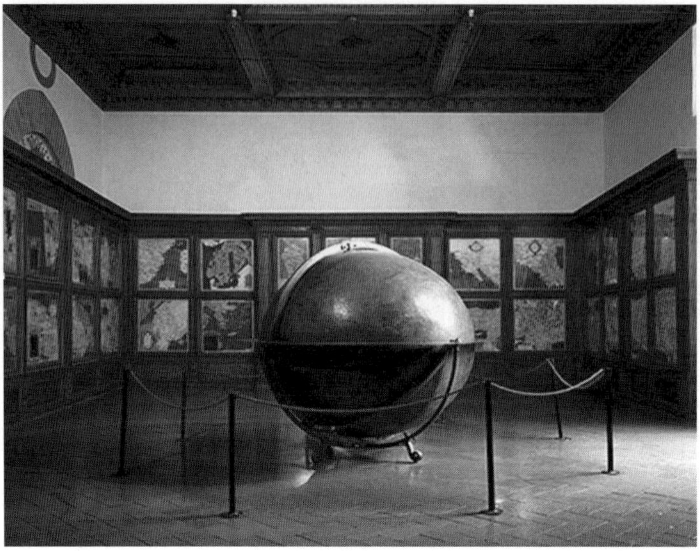

**Fig. 61.** The Map Room ("Sala delle Carte") with the old globe, in Palazzo Vecchio (Florence).

also involved an exploration of the nature and condition of its inner structure. It was therefore decided that in addition to the surface diagnosis, a CT scan would be performed. Our Department was assigned the task of performing this inspection. The difficulties of performing an *in-situ* CT scan of such a large object (maybe the largest ever subject to CT *in situ*) in a museum surrounded by visitors, were immediately evident. It was therefore necessary to perform the scan by night. If the "cone-beam" mode was chosen, the projection of the globe would ideally be on a screen with a surface area of $4 \times 4$ m$^2$, located about 5 m from the X-ray source. With a planar detector, with dimensions of $30 \times 40$ cm$^2$, about 33 000 images would be taken. In order to test the feasibility of performing the measurements, it was decided to take preliminary digital radiographs using a new type of fast EBCCD camera. This appliance, developed by the Russian firm Geosphaera, has a CCD with $528 \times 286$ pixels, a read-out time of 25 ms and a dynamic range of 12 bits. As the preliminary radiographs, taken in July 2003, gave good results, it was decided to proceed using the same system. The acquisition time, for one image, was of the order of 5 s. As a comparison, a normal radiograph using film took about 20 min for a distance of less than 3 m. In the detection system, shown in Fig. 62, the camera looks directly at the GOS scintillating screen; the 45° mirror is not necessary, as the X-ray intensity impinging on the camera at that distance (5 m) is very low. The globe (weighing about 1000 kg) was placed on a rotating platform. The set-up of the whole system is shown in Fig. 63. A motor moved the X-ray tube along an aluminium column. Two more motors moved the detector along the $x$- and $y$-axis.

In the case of the globe, it was not possible to perform CT in "cone beam" mode. In fact, taking into account the rule formulated in Section 8.2.5, (the angle of the cone must be

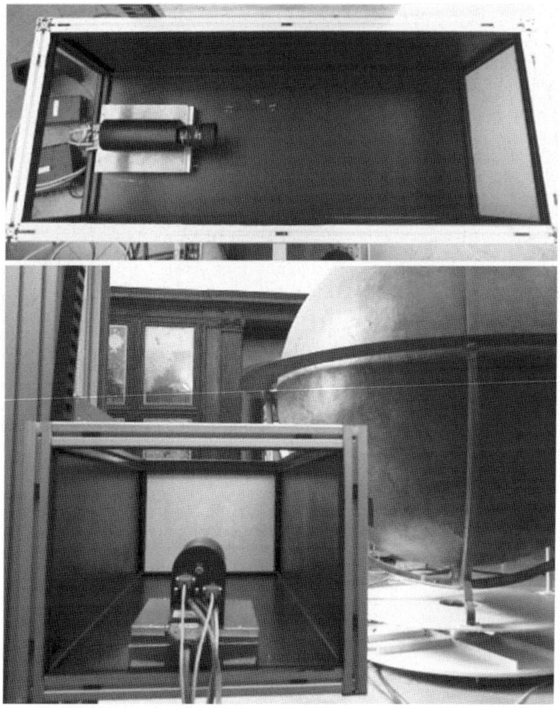

**Fig. 62.** The intensified TVcamera inside the box.

smaller than 10°), the X-ray source would be located too far from the detector. It was therefore decided to put the X-ray tube at different heights and to move the detector, aligned with the tube, in a horizontal direction only. In this way, we had 14 "cone beam" CTs but each one with a small angle, thus making it possible to apply the FDK approximation correctly.

To minimise the X-rays in the room and in the part of the globe not involved, the tube was equipped with a lead collimator. The preliminary radiographs showed that the internal structure was made of iron as reported in ancient documents written by Egnazio Danti. It was therefore necessary to adopt a portable 200 kV system (manufactured by Gilardoni S.P.A.). In order to pass through the iron structure and to minimise the artefacts, a voltage of 180 kV was used.

In short, the CT of this large globe was obtained thus:

(a) the X-ray tube was placed on the North Pole;
(b) the detector, located on the horizontal trail as shown in Fig. 63, was placed to the extreme right of the projection of the globe. At this point, an image was acquired;
(c) the platform, on which the globe was placed, was rotated in angular steps of 1°. After 360 acquisitions, the detector was translated by about 40 cm, then another 360 acquisitions were performed and so on, until the entire slice had been scanned;

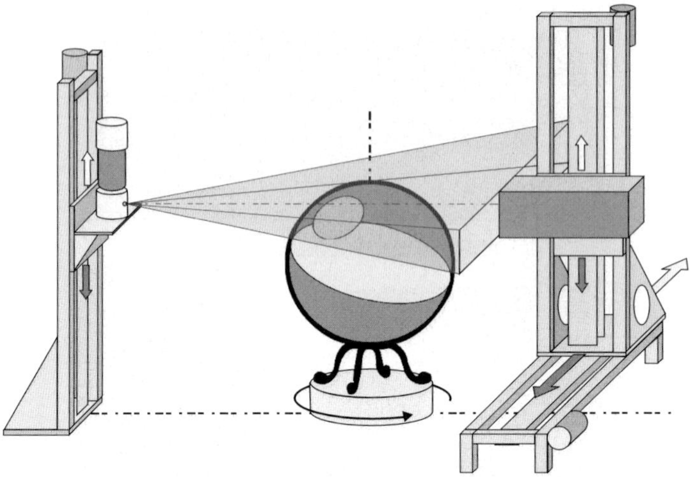

**Fig. 63.** Sketch of the system. The globe rotates over a platform with angular steps of 1°.

(d) the tube was moved towards the South Pole and the operation was repeated, starting from point (b).

Each image was identified by three figures for slice, position of detector and angle. For a defined angle, these images were "welded" as shown in Fig. 64, corresponding to the slice No. 7 from the top. Figure 65 shows the different slices together. At this point, the 3D reconstruction was performed using appropriate software developed in our Department.

The inner structure, made of iron, then appeared as shown in Figs. 66 and 67. It was performed using a central pole, 8 bars as 2 tetrahedrons and 30 meridians. Using the segmentation of the image (see Section 4.3.3), it was possible to evaluate the volume of the iron inside, which weighed about 350 kg.

Ancient documents report that several pounds of hemp had been bought but it is not sure that the hemp was used in creating the globe. Probably it was put between the surface and the iron structure as shown by Fig. 68. The image is very noisy as the hemp is rather transparent to X-rays of the energy used. The measurements took about one month (June 2004) to complete and were performed by young researchers and PhD students (in Physics and Computer Science) at the University of Bologna (Fig. 69).

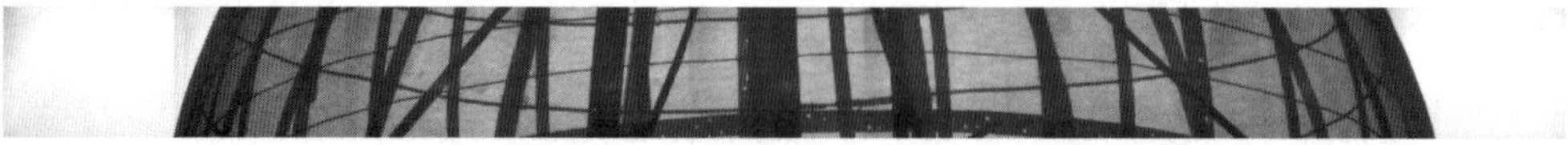

**Fig. 64.** One of the fourteen slices obtained.

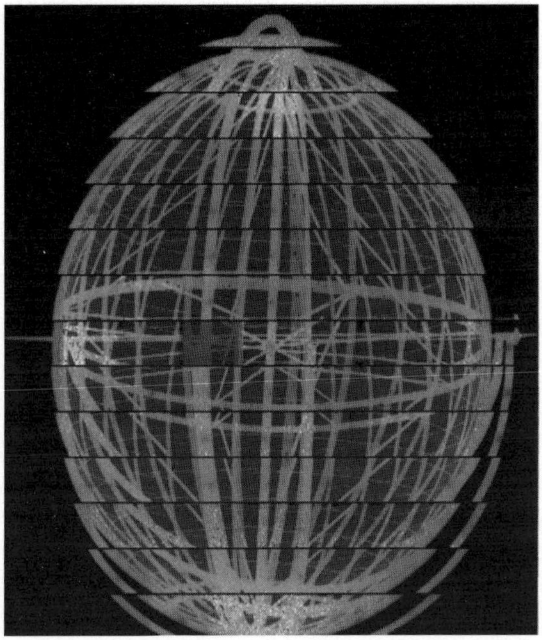

**Fig. 65.** All 14 slices together.

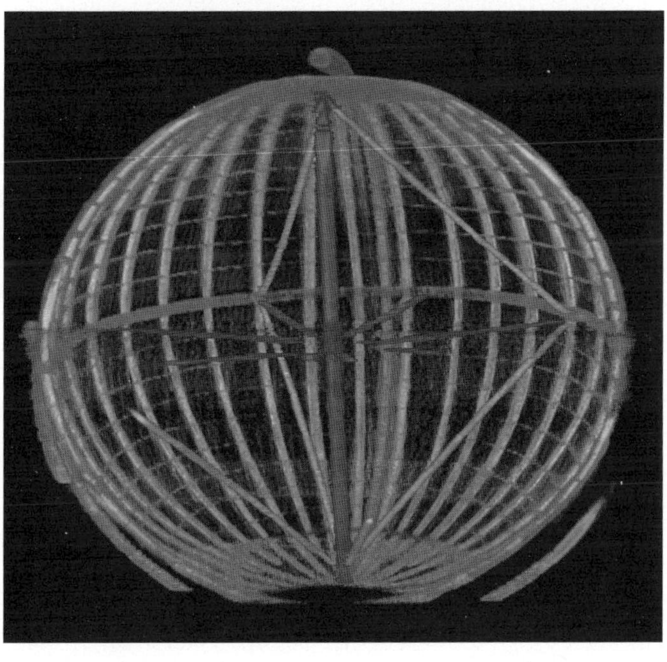

**Fig. 66.** A three-dimensional reconstruction.

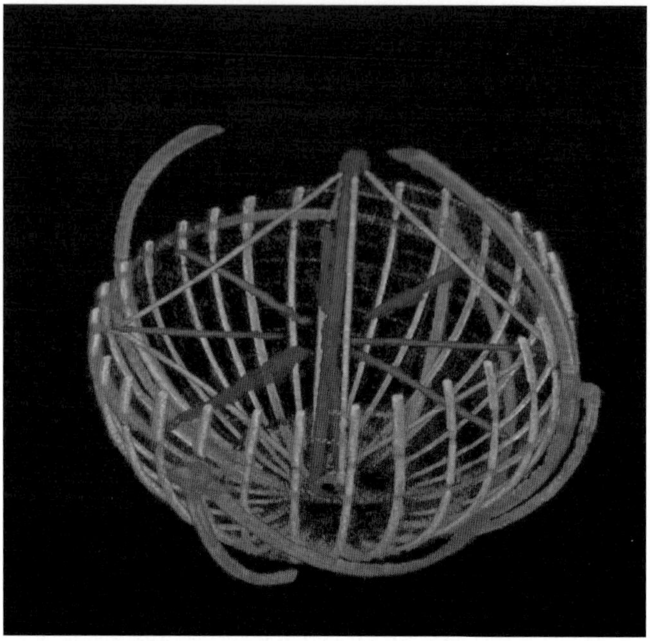

**Fig. 67.** An exploded 3D reconstruction.

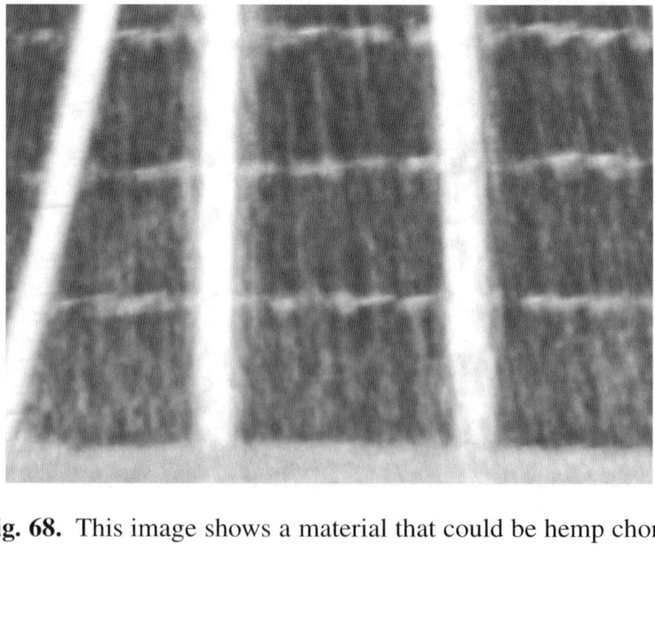

**Fig. 68.** This image shows a material that could be hemp chord.

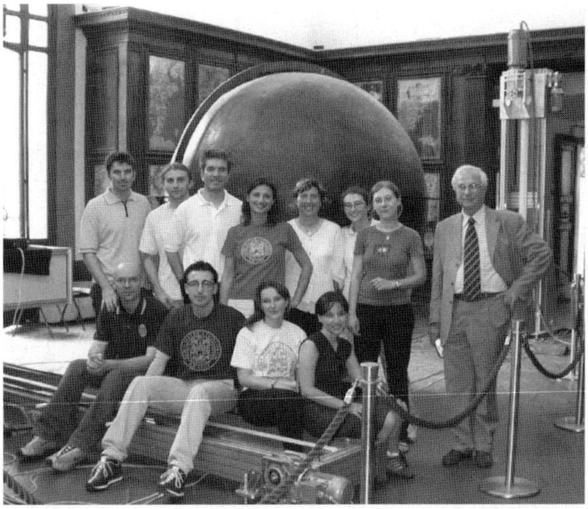

**Fig. 69.** The team of young researchers and PhD students (physicists and computer scientists) with the author, at the end of the measurement phase. From left to right, they are: (stand up) Alessandro Pasini, Nico Lanconelli, Matteo Bettuzzi, Samantha Cornacchia, Maria Pia Morigi, Marilisa Giordano, Alice Miceli, the author; (sit down) Alessandro Fabbri, Davide Bianconi, Carlotta Cucchi, Emilia di Nicola, Not in picture: Davide Romani, Alberto Rossi, Rossella Brancaccio.

## 9.5. Neutron tomography

Generally speaking, neutron tomography is performed using thermal neutron beams produced by nuclear reactors [29] or by suitable facilities [42]. Less frequently, cold [43] and fast neutrons are also used [44].

In general, the beams have a circular section and are of the parallel type (like synchrotron light). Acquisition is performed by planar detectors based on $^{6}Li$ and ZnS [28]. Neutron DRs and CTs of a model of a small helmet and of an ancient amulet (cat) are reported in Fig. 70 [45]. Where possible, it is very interesting to compare (or overlap) DRs and CTs performed using X-rays (or $\gamma$ rays) and those using neutrons [46].

## 9.6. Induced activation by X-rays and neutrons

### *9.6.1. Activation by X-rays*

When high-energy photons interact with atomic nuclei, there is the possibility that the struck nucleus will emit a neutron that, in turn, makes the surrounding materials radioactive. Each nucleus has a threshold energy for this photoreaction, below which the neutron is not emitted. The smallest threshold energy is 1.6 MeV, for $^{9}Be$, then 2.2 MeV for $^{2}H$, therefore if one uses radioisotopic sources like $^{60}Co$ or $^{135}Cs$, which emit photons with lower energies, it is physically impossible for the irradiated materials to become radioactive.

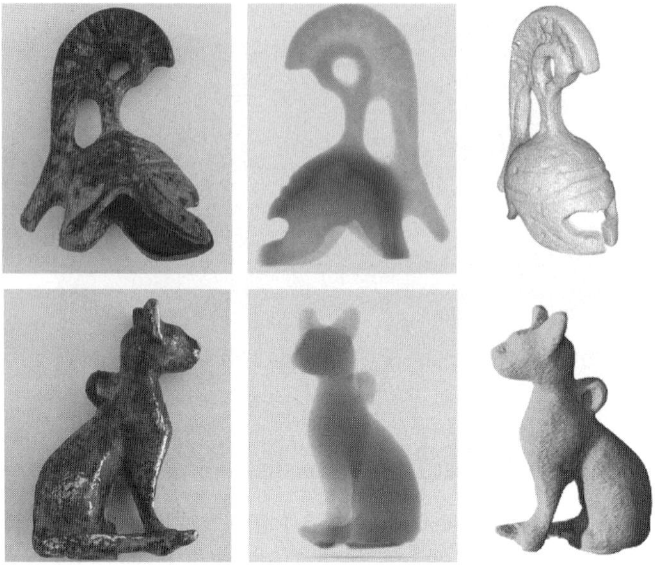

**Fig. 70.** Photos, neutron radiographies and neutron tomographies of a model of a small helmet and of an ancient amulet (cat).

On the contrary, if one uses LINACs as a photon source (see Section 2.3.2), the emission of neutrons is possible, especially if heavy materials, like lead, are irradiated. The threshold energy for Pb is about 7 MeV. In practice, for LINACs with energy less than 10 MeV, induced radioactivity is rather small, almost negligible.

*9.6.2. Activation by neutrons*

Apart from the difficulty of obtaining neutrons, the main problem of neutron radiography or tomography is that the sample becomes radioactive. The induced radioactivity can be so high that for some tests (*e.g.* real time radioscopy with high neutron fluxes), the sample cannot be handled for many days or months.

**Warning!** When high energy LINACs or neutrons are used as radiation source, always refer to an expert in radiation protection.

This recommendation is also valid when "in the field" measurements are performed, whatever the source.

## 10. SUGGESTIONS AND CONCLUSIONS

An increase in scientists' interest in cultural heritage and decrease in humanists' suspicion of technology have overcome an age-old debate in understanding the works of art. After all, were not Leonardo and Michelangelo, both scientists and artists? Many new applications based on physical techniques are being developed. In this chapter, we have provided some basic information on digital radiography and computed tomography without the presumption to be exhaustive. It should however be pointed out that the CT field is a very difficult one. The ease with which CT can be performed in the medical field may prove deceptive: medical CT was designed for the human body (composed mainly of water) alone and cannot be successfully used on bodies with different shapes or compositions. In order to perform good, non-destructive evaluations, the most suitable DR or CT system (source, moving equipment, detector and elaboration software) must be carefully chosen to avoid obtaining disappointing results, wasting time and money and … losing faith in Physics.

## APPENDIX A: BASIC NOTIONS CONCERNING FOURIER TRANSFORMS

### A.1. The Fourier series

This appendix contains some of the basic concepts used in the imaging field.

The Fourier transform can be considered as an extension of the development in Fourier series that, for a periodic function of period $T$, has the following expression:

$$f(t) = \frac{1}{2}a_0 + \sum_{n=1}^{\infty}\left(a_n \cos\frac{2\pi}{T}nt + b_n \sin\frac{2\pi}{T}nt\right) \tag{A.1}$$

where $a_0$, $a_n$ and $b_n$ are expressed by:

$$\begin{aligned} a_0 &= \frac{2}{T}\int_{-T/2}^{T/2} f(t)\mathrm{d}t, \\ a_n &= \frac{2}{T}\int_{-T/2}^{T/2} f(t)\cos\frac{2\pi}{T}nt\mathrm{d}t, \\ b_n &= \frac{2}{T}\int_{-T/2}^{T/2} f(t)\sin\frac{2\pi}{T}nt\mathrm{d}t. \end{aligned} \tag{A.2}$$

If $f(t)$ is a symmetric function, only the terms $a_0$ and $a_n$ are not equal to zero.

Figure A.1(A) shows a sinusoidal signal affected by noise. By developing this function into a Fourier series and taking the fundamental harmonic only, the noise, that is the remaining part of the series, is removed (Fig. A.1(B)). Conversely, by taking all harmonics with $n > 1$, one retains the noise and discards the regular shape of the signal (Fig. A.1(C)).

With a simple algebraic operation, making use of Euler's formula:

$$e^{jt} = \cos t + j\sin t, \tag{A.3}$$

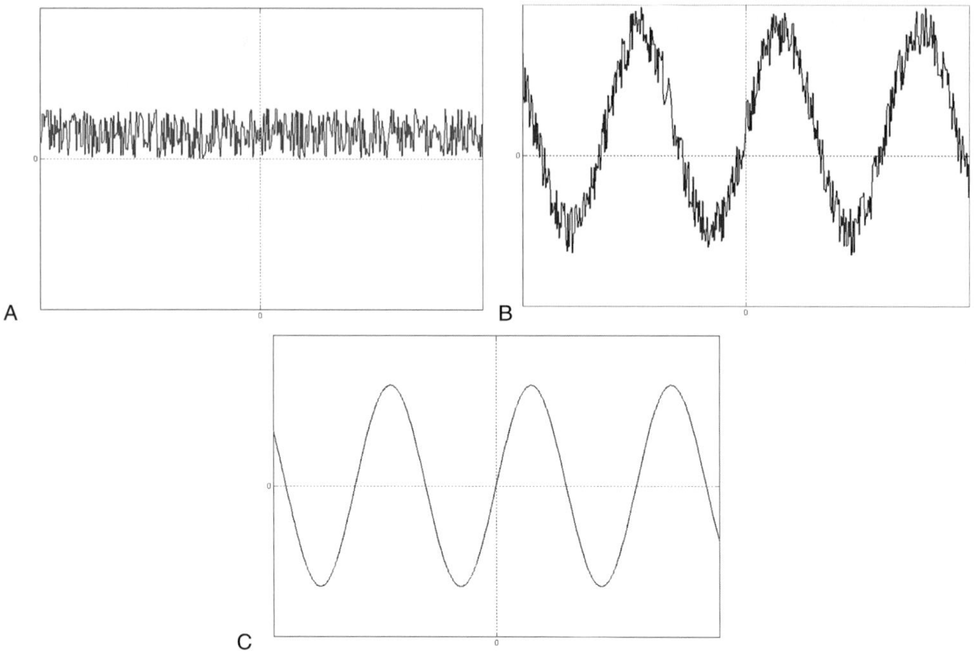


**Fig. A.1.** (A) A sinusoidal function with a superimposed noise; (B) only fundamental harmonic is retained; (C) the noise (difference between the whole signal and the fundamental harmonic).

the equations (A.1) and (A.2) assume the form:

$$f(t) = \sum_{n=-\infty}^{+\infty} c_n \exp\left( j\frac{2\pi}{T}nt \right), \tag{A.4}$$

where:

$$c_n = \frac{1}{T}\int_{t_0}^{t_0+T} f(t)\exp\left( -j\frac{2\pi}{T}nt \right)\mathrm{d}t. \tag{A.5}$$

One important relation is:

$$\frac{1}{2\pi}\int_{-\pi}^{+\pi} \{f(t)\}^2\mathrm{d}t = \sum_{-\infty}^{+\infty}|c_n|^2.$$

The totality of $a_0$, $a_n$, $b_n$ or $c_n$ where, $n = 1, 2, \ldots, \infty$, is defined as the "spectrum" of that function. Giving the infinite values constituting the spectrum is equivalent to giving the infinite values of the function $f(t)$ for each point of the interval $(0,T)$.

## A.2. One-dimensional Fourier transform

It is possible to demonstrate that, by extending the integration limits from $-\infty$ to $+\infty$ and proceeding as in the discretised case, equations (A.4) and (A.5) are transformed into:

$$f(x) = \int_{-\infty}^{+\infty} F(u)e^{j2\pi ux}\mathrm{d}u, \tag{A.6}$$

$$F(u) = \int_{-\infty}^{+\infty} f(x)e^{-j2\pi ux}\mathrm{d}x. \tag{A.7}$$

$F(u)$ is known as the *Fourier Transform* of $f(x)$ and is equivalent to the aforementioned spectrum. The two equations comprise the *Fourier Transform pair*.

$f(x)$, which can be obtained by $F(u)$, is also known as the *inverse Fourier Transform*.

In general, the transformed function is a complex function, for which one can use the usual notation for complex numbers:

$$F(u) = R(u) + jI(u), \tag{A.8}$$

or in the exponential form, making use of Euler's relation:

$$F(u) = |F(u)| \cdot e^{j\Phi(u)}, \tag{A.9}$$

where:

$$|F(u)| = \{R(u)^2 + I(u)^2\}^{1/2} \tag{A.10}$$

is denominated the magnitude or spectrum of the Fourier Transform and

$$\Phi(u) = \tan^{-1}\frac{I(u)}{R(u)} \tag{A.11}$$

is the *phase angle.*

The square of the modulus:

$$E(u) = |F(u)|^2 = R(u)^2 + I(u)^2 \tag{A.12}$$

is known as the energy spectrum or power spectrum of $f(x)$.

A *Fourier pair*, with

$$f(x) = \begin{cases} a & 0 \le x \le x_0 \\ 0 & \text{otherwise} \end{cases}$$

is shown in Fig. A.2. It should be remembered that $I(u)$, $R(u)$ and $|F(u)|$ extend to infinity, even if $f(x)$ differs from zero in a finite interval.

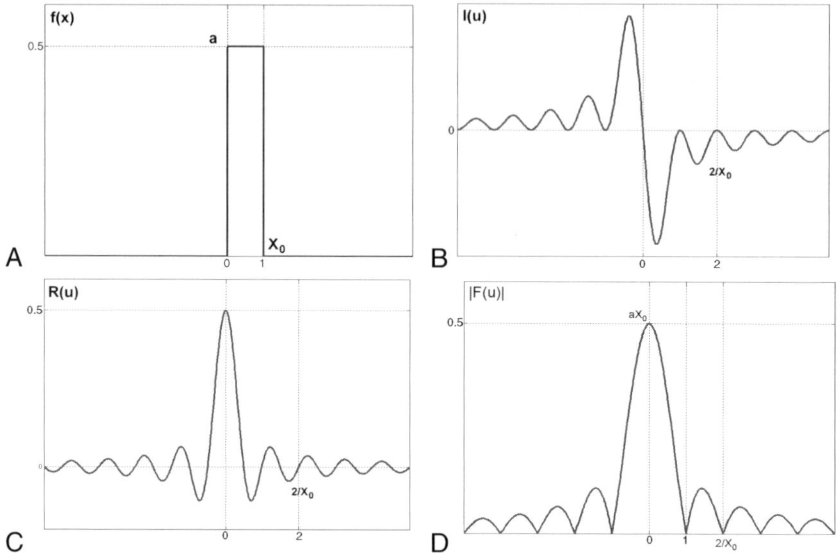


**Fig. A.2.** (A) $f(x)$, Rectangular function; (B) $I(u)$, imaginary component of $F(u)$; (C) $R(u)$, real component of $F(u)$; (D) spectrum of $F(u)$.

### A.3. Two-dimensional Fourier transform

The extension to two variables, $u$ and $v$, gives:

$$F(u,v) = \int_{-\infty}^{+\infty}\int_{-\infty}^{+\infty} f(x,y)\mathrm{e}^{-j2\pi(ux+vy)}\mathrm{d}x\mathrm{d}y, \tag{A.13}$$

for the transform and

$$f(x,y) = \int_{-\infty}^{+\infty}\int_{-\infty}^{+\infty} F(u,v)\mathrm{e}^{j2\pi(ux+vy)}\mathrm{d}u\mathrm{d}v, \tag{A.14}$$

for the inverse transform.

Once again, in two-dimensional cases, the transformed function is a complex one. Using the notation for complex numbers:

$$F(u,v) = R(u,v) + jI(u,v), \tag{A.15}$$

or, making use of Euler's relation in exponential form:

$$F(u,v) = \left|F(u,v)\right| \cdot \exp(j\Phi(u,v)), \tag{A.16}$$

where:

$$|F(u,v)| = \{R(u,v)^2 + I(u,v)^2\}^{1/2}, \tag{A.17}$$

and

$$\Phi(u,v) = \tan^{-1}\frac{I(u,v)}{R(u,v)}, \tag{A.18}$$

with the same notation as in the mono-dimensional case.

## A.4. One-dimensional discrete Fourier transform

Because we are interested in processing digital images, which are equivalent to numerical matrices, equations (A.13) and (A.14) must be rewritten for discrete functions [18].

Starting from a one-dimensional discrete function, $f(x)$, $x = 0, 1, 2, \ldots N-1$, the *discrete Fourier Transform* (DFT) is:

$$F(l) = \frac{1}{N}\sum_{n=0}^{N-1} f(n)\exp\left[-j2\pi\frac{nl}{N}\right], \quad l = 0,\ 1, \ldots,\ N-1 \tag{A.19}$$

computed for values of $u = 0, 1, 2, \ldots, N-1$.

As we are dealing with digital images, we have assumed the sampling intervals to be constant, and because we started with a sampled function, its transform is also sampled. The definition intervals have been properly normalised so that the total interval is equal to 1.

The inverse transform, is:

$$f(k) = \sum_{l=0}^{N-1} F(l)\exp\left[\frac{j2\pi kl}{N}\right], \quad k = 0,\ 1, \ldots,\ N-1 \tag{A.20}$$

The correspondence of the two transforms is immediately demonstrated by substituting (A.20) with (A.19) or *vice versa.*

## A.5. Two-dimensional discrete Fourier transform

The extension of DFT in two-dimensions is fairly straightforward. The DFT of a two-dimensional function $f(x, y)$ (far more interesting with regard to digital image processing) with a size of $M \times N$ is given by the equation:

$$F(k,l) = \frac{1}{NM}\sum_{m=0}^{M-1}\sum_{n=0}^{N-1} f(m,n)\exp\left[-j2\pi\left(\frac{km}{M} + \frac{ln}{N}\right)\right], \tag{A.21}$$

$k = 0,\ 1, \ldots,\ M-1;\ l = 0,\ 1, \ldots,\ N-1.$

Similarly:

$$f(m,n) = \sum_{k=0}^{M-1}\sum_{l=0}^{N-1} F(k,l)\exp\left[j2\pi\left(\frac{km}{M}+\frac{ln}{N}\right)\right], \tag{A.22}$$

$m = 0, 1, \ldots, M-1;\ n = 0, 1, \ldots, N-1$

Equations (A.21) and (A.22) are written for rectangular matrices $M \times N$, and comprise the *two-dimensional, discrete Fourier Transform (DFT) pair.*

For square matrices $N \times N$, these equations can be rewritten as:

$$F(k,l) = \frac{1}{N}\sum_{m=0}^{N-1}\sum_{n=0}^{N-1} f(m,n)\ \exp\left[-j2\pi\left(\frac{km+ln}{N}\right)\right], \tag{A.23}$$

$k = 0, 1,\ldots, N-1;\ l = 0, 1,\ldots, N-1$

Similarly:

$$f(m,n) = \sum_{k=0}^{N-1}\sum_{l=0}^{N-1} F(k,l)\exp\left[j2\pi\left(\frac{km+ln}{N}\right)\right], \tag{A.24}$$

$m = 0, 1,\ldots, N-1;\ n = 0, 1,\ldots, N-1$

The calculation for the spectrum, the phase and the energy spectrum is performed as seen in the continuum.

If $N$ is a power of 2 (*i.e.* $N = 2^n$), the calculation time of $F(k, j)$ and $f(m, n)$ is drastically reduced (of a factor $N/\log_2 N$) by using an algorithm known as the *Fast Fourier Transform* (FFT) [17]. The FFT algorithm is normally used in digital image processing.

## A.6. Some properties of 2D discrete Fourier transforms

*Mean value*

Let us consider the discrete function $f(x, y)$. Its mean value is given by

$$\bar{f} = \frac{1}{N^2}\sum_{m=0}^{N-1}\sum_{n=0}^{N-1} f(m,n). \tag{A.25}$$

Moreover, for $k = l = 0$, equation (A.23) becomes

$$F(0,0) = \frac{1}{N}\sum_{m=0}^{N-1}\sum_{n=0}^{N-1} f(m,n). \tag{A.26}$$

We can conclude that

$$\bar{f} = \frac{1}{N} F(0,0) \tag{A.27}$$

which shows us that the term $F(0,0)$ corresponds to $N$-times the average grey level of the image.

*Periodicity and symmetry*

From equation (A.23), one can easily show that:

$$F(k,l) = F(k+N,l) = F(k,l+N) = F(k+N,l+N) \tag{A.28}$$

The same occurs for $f(x, y)$ in the spatial domain. This peculiarity highlights the fact that the *Fourier Transform* and its inverse repeats indefinitely in both spatial dimensions.

It is not difficult to demonstrate that

$$[f(x, y)(-1)^{x+y}] = F(u - N/2, v - N/2) \tag{A.29}$$

This relation states that the origin of the *Fourier Transform* of $f(x,y)(-1)^{x+y}$, that is $F(0,0)$, is at $u = N/2$ and $v = N/2$. The translation of the origin gives a better representation of $F(u,v)$, as shown in Fig. A.3.

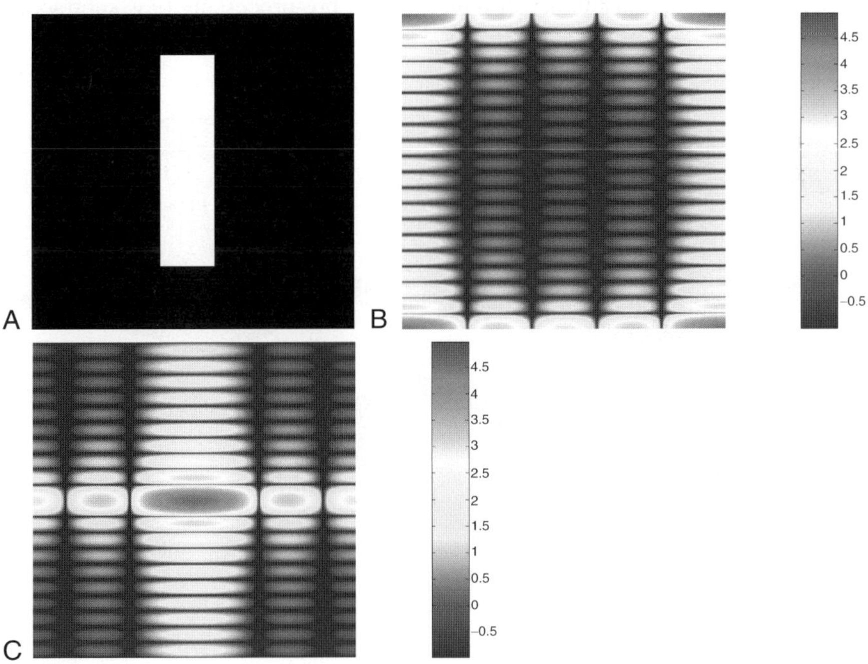


**Fig. A.3.** (A) Image; (B) *Fourier Transform* of (A) with the origin in (0,0); (C) *Fourier Transform* of (A) with origin in ($N$/2, $N$/2).

There are also other interesting properties, such as:

$$F(k,l) = F^*(-k, -l), \tag{A.30}$$

where $F^*$ is the complex conjugate of $F$. Passing to absolute values:

$$|F(k,l)| = |F(-k, -l)|. \tag{A.31}$$

## A.7. Filtering in the frequency domain

Figure A.1 shows that, given a signal and its spectrum, the slowly varying components of the signal are related to the low frequencies in the spectrum. Conversely, the rapidly varying components (*e.g.* the "noise") refer to the high frequencies. By filtering the high frequencies, we can "clean" the signal. The same happens for an image (signal) and its *Fourier Transform* (spectrum). We can decrease the noise by decreasing the importance of the high frequencies (*lowpass filtering*) or enhance the contrast by decreasing the importance of the low frequencies (*highpass filtering*). The filtering is performed by multiplying $F(u, v)$ by a function $H(u, v)$, named *filter transfer function*, which decreases certain frequencies and leaves the others unchanged. Some shapes of $H(u, v)$ are shown in Fig. A.4.

Fourier domain filtering is performed as follows:

(a) acquire the image $f(x, y)$ and perform all possible pre-processing (*e.g.* subtract noise);

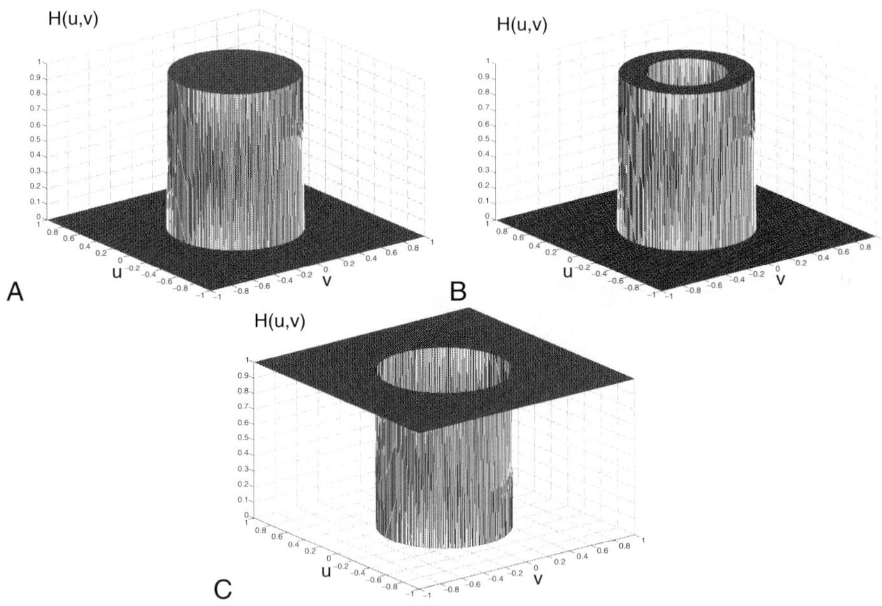


**Fig. A.4.** Some shapes of the filter function $H(u,v)$. (A) "Ideal" low-pass filter; (B)"Ideal" selecting band filter; (C) "Ideal" high-pass filter.

(b) centre $F(u, v)$, multiplying $f(x, y)$ by $(-1)^{x+y}$, as indicated in equation (A.29);
(c) compute $F(u, v)$ of the image;
(d) multiply $F(u, v)$ by the filter function $H(u, v)$;
(e) compute the inverse DFT of $G(u, v) = H(u, v)\, F(u, v)$;
(f) multiply the real part of the Filtered Image $F^{-1}[G(u, v)]$ by $(-1)^{x+y}$.

It should be pointed out that with this procedure we enhance the image as a whole; it is not a "local treatment".

### A.8. Convolution of two functions

The convolution of two functions is a very important operation for image restoration. The "local treatment" can be traced back to a convolution of a function (the *image*) with another function (the *mask* or the *filter*).

The convolution of two functions $f(x)$ and $h(x)$, formally stated as $f(x) * h(x)$, is defined by the integral:

$$g(x) = f(x) * h(x) = \int_{-\infty}^{+\infty} f(\xi)h(x-\xi)\mathrm{d}\xi. \tag{A.32}$$

One important case is when $h(x)$ is the delta function $\delta(x)$ (Dirac function or pulse function), which has the following properties:

$$\int_{-\infty}^{+\infty} f(x)\delta(x-x_0)\mathrm{d}x = f(x_0), \tag{A.33a}$$

$$\int_{-\infty}^{+\infty} \delta(x-x_0)\mathrm{d}x = \int_{x_0-\varepsilon}^{x_0+\varepsilon} \delta(x-x_0)\mathrm{d}x = 1. \tag{A.33b}$$

Using equation (A.26), it is easy to demonstrate that the convolution of any function with a delta function gives a function that is a translated copy of the original function. The extension to two-dimensional functions is straightforward.

*Convolution Theorem*

The main importance of the convolution operation is connected to the "Convolution Theorem". If $f(x, y)$ and $h(x, y)$ have the functions $F(u, v)$ and $H(u, v)$ as Fourier transforms respectively, the first part of the convolution theorem states that $f(x,y) * h(x,y)$ has the function $F(u, v) \cdot H(u, v)$ as a Fourier transform. These results can be formally presented as:

$$f(x, y) * h(x, y) \Leftrightarrow F(u, v) \cdot H(u, v). \tag{A.34}$$

The double arrows ($\Leftrightarrow$) indicate that the convolution of $f(x, y)$ with $h(x, y)$ can be obtained by the Fourier transforms of $f(x, y)$ and, $h(x, y)$, performing the multiplication element-by-element (or pixel by pixel) of $F$ and $H$, then taking the inverse Fourier transform of the function-product obtained.

Calculating a convolution integral using the procedure outlined above is much faster than a direct calculation.

Equation (A.34) can also be written in the form:

$$f(x, y) \cdot h(x, y) \Leftrightarrow F(u, v) * H(u, v), \tag{A.35}$$

which states that the convolution in the frequency domain is equivalent to a multiplication in the spatial domain.

## APPENDIX B: MODULATION TRANSFER FUNCTION

### B.1. Point spread function, line spread function and edge spread function

Another important application of the Fourier transform is the determination of the modulation transfer function (MTF). The MTF is the principal function that quantifies the spatial resolution of an image-acquiring system.

Let us perform a simple experiment. Take a well-focused overhead projector and place a piece of cardboard with a sharp edge over it. If we now look at the projected image, we realise that it has a black-and-white distribution, like a "step function", as shown in Fig. B.1(A). If we move the cardboard away, the edge becomes less sharp; it will no longer be a step function but something similar to that shown in Fig. B.1(B).

There will be a "spread" on the boundary: we have produced an "edge spread function" (ESF). The blurring will depend on the distance of the cardboard from the overhead

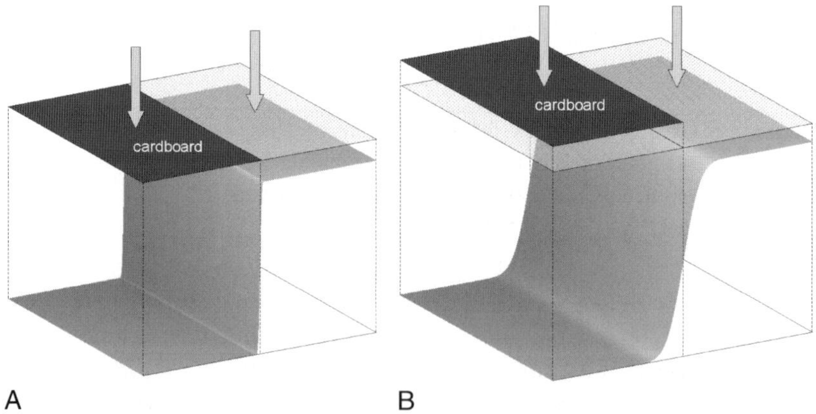


**Fig. B.1.** A cardboard, with a sharp edge, is put over a well-focused overhead projector plane. The image produced is a black-and-white image as a "step function" (curve (A)). The cardboard is then moved away and the grey distribution is similar to a step function but not so sharp like in the former case (curve (B)). The step function is affected by "blurring".

projector plane. The same will happen if we have a small slit in the cardboard. When the entire system is well focused, the grey distribution produced is like a narrow "rectangular function", which can be approximated with a mono-dimensional Dirac delta function, $\delta(x-x_0)$. By moving the cardboard away as described above, we obtain the image shown in Fig. B.2 (like a Gaussian). We have done the so-called "line spread function" (LSF). As an extension of this, a small hole can be represented as a 2D delta function, $\delta(r-r_0)$, and the unfocused one will be the "point spread function" (PSF) (Fig. B.3). This means that an input "point signal" undergoes a sort of 2D gaussian blurring, given by the shape (width) of the PSF.

Now let us consider a well-collimated linear beam of photons or neutrons, such as that obtained from a synchrotron or collimator in a nuclear reactor. Let these particles pass through a thin slit, which we consider as $\delta(x-x_0)$, and impinge on a scintillating screen. A flash of light will be produced isotropically in the point where an interaction with the scintillating material occurs: for instance, at a depth $z$ from the screen face where the image is formed. The light will escape the screen within a broad angle so that the spatial distribution of the escaped light, $L(x-x_0)$,[8] will have the shape of an LSF. The nearer the image plane, the sharper the LSF. It is possible to demonstrate that [47]:

$$L(x-x_0) = \left[\frac{\lambda/\pi}{1+\lambda^2(x-x_0)^2}\right], \tag{B.1}$$

where $\lambda = z^{-1}$ ($\lambda$ is called "resolution parameter").

The maximum of this function occurs for $x = x_0$

$$L(x = x_0) = \frac{\lambda}{\pi}. \tag{B.2}$$

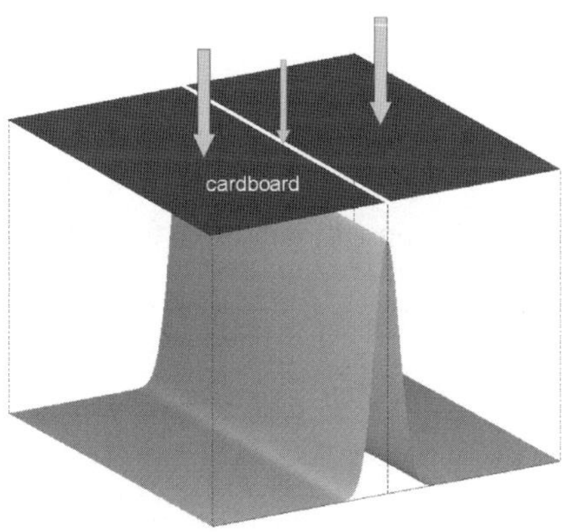


**Fig. B.2.** A cardboard, with a thin slit, is put over a well-focused overhead projector plane. The grey distribution is a narrow "rectangular function", like a mono-dimensional Dirac delta function. The cardboard is then moved away and the grey distribution becomes similar to a Gaussian. We have obtained the so-called "line spread function".

[8] The right term for equation B.1 is the so-called "Lorentzian" function (see Mathworld.walfram.com) and, in general, it appears when resonance phenomena are dealt with.

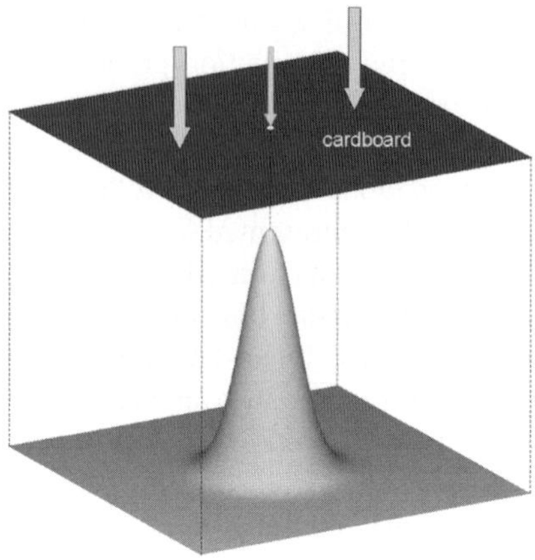


**Fig. B.3.** A cardboard, with a small hole, is put over a well-focused overhead projector plane. The grey distribution is like a two-dimensional Dirac delta function. The cardboard is then moved away and the grey distribution becomes similar to a Gaussian. Thus we have obtained the so-called "point spread function".

The width at which $L(x-x_0)$ assumes half of its maximum value is defined as "full width at half maximum" (FWHM) (see Fig. B.4)

$$\frac{1}{2}\left(\frac{\lambda}{\pi}\right) = \frac{\lambda/\pi}{1+\lambda^2 x_{1/2}^2} \tag{B.3}$$

so that the

$$\text{FWHM} = \frac{2}{\lambda} \tag{B.4}$$

Low FWHM means high resolution and therefore the FWHM is sometimes assumed as an index of the spatial resolution of the system.[9]

[9] As $x_0$ is an arbitrary value, we can put $x_0 = 0$, and equation (B.1) becomes:

$$L(x) = \left[\frac{\lambda/\pi}{1+\lambda^2 x^2}\right] \tag{i}$$

When other factors exist that degrade the LSF (such as background and noise), they are combined and then equation (i) is assumed to be of a gaussian type:

$$L(x) = \frac{\alpha}{\sqrt{\pi}}\exp\left(-\alpha^2 x^2\right), \tag{ii}$$

where $\alpha$ is called the "resolution parameter".

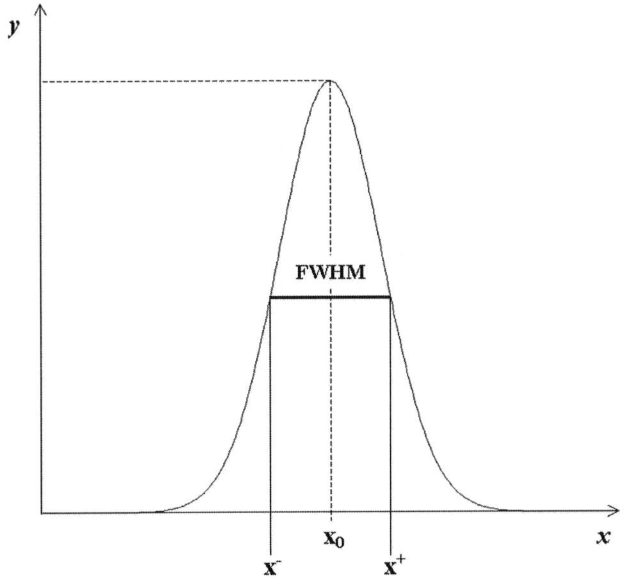


**Fig. B.4.** At the values $x^-$ and $x^+$, the function has half of its maximum value. The Full Width at Half Maximum (FWHM), $(x^+ - x^-)$ is equal to $2/\lambda$, an index of the spatial resolution of the system.

The edge spread function, $S(x)$, can be intended as a superposition of infinite line spread functions. This can be obtained by integrating $L(x - x_0)$ up to $+\infty$:

$$S(x) = N\int_0^\infty L(x - x_0)\mathrm{d}x_0, \tag{B.5}$$

where $N$ is a normalisation parameter.

By substituting $L(x - x_0)$ with its expression, given by equation B.1, we have:

$$S(x) = N\int_0^\infty \left(\frac{\lambda/\pi}{1 + \lambda^2(x - x_0)}\right)\mathrm{d}x_0. \tag{B.6}$$

With the position $t = (x_0 - x)\lambda$, the integral assumes the following shape:

$$S(x) = \frac{N}{\pi}\int_{-\lambda x}^{+\infty}\left(\frac{\mathrm{d}t}{1 + t^2}\right) = \frac{N}{\pi}[\mathrm{arctg}(t)]_{-\lambda x}^{+\infty} = \frac{N}{\pi}\left[\frac{\pi}{2} - \mathrm{arctg}(-\lambda x)\right], \tag{B.7}$$

which, finally, gives the edge spread function (ESF):

$$S(x) = N\left[\frac{1}{2} + \frac{1}{\pi}\text{arctg}(\lambda x)\right]. \tag{B.8}$$

Conversely, the differentiation of the ESF gives the LSF. As it is easier to measure the ESF than the LSF, we will obtain the MTF by measuring the ESF, as demonstrate below.

For instance, if the measured function is the optical density of a film, $D(x)$, it can be represented as the superposition of the real signal, noise and the background (see Fig. B.5):

$$D(x) = D_0(x) + D_n + D_b \tag{B.9}$$

where:

$D_0(x)$ = component associated with the recording system;

$D_n$ = component associated with the statistical noise (*e.g.* granular composition of the film);

$D_b$ = optical density associated with the film not directly exposed to the radiation.

It is possible to infer the $\lambda$ of the system by fitting the function $D(x)$ by the expression on the right hard side of equation (B.8). The FWHM is then obtained by equation (B.4).

Another simpler way to obtain the FWHM is as given below:

For $x = \lambda^{-1}$, we get

$$D(\lambda^{-1}) = D(\lambda^{-1}) = \frac{1}{2} + \frac{1}{\pi}\text{arctg}(\lambda\lambda^{-1}) = \frac{1}{2} + \frac{1}{\pi}\frac{\pi}{4} = \frac{3}{4} = 0.75,$$

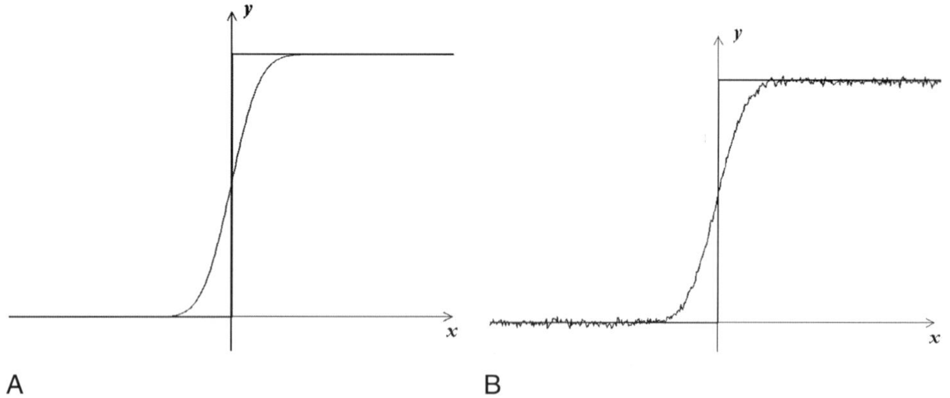


**Fig. B.5.** The measured distribution of an ESF affected by noise.

and for $x = -\lambda^{-1}$

$$D(-\lambda^{-1}) = \frac{1}{2} - \frac{1}{\pi}\text{arctg}(-1) = \frac{1}{2} - \frac{1}{4} = \frac{3}{4} = 0.25,$$

so that, having measured the ESF, its 75% corresponds to $x_1 = 1/\lambda$ and its 25% corresponds to $x_2 = -1/\lambda$. Their difference, $(x_1 - x_2) = 2/\lambda = \text{FWHM}$.

We can conclude that by measuring ESF, it is possible to obtain the FWHM related to the spatial resolution of the acquisition system.

## B.2. Optical Transfer Function and Modulation Transfer Function

*Introduction*

If we have a "step function" as the "input function" of a system, we will have a smoother curve as the "output function". In other words, if we have an input function $f_i(x_0)$, given by:

$$f_i(x_0) = \begin{cases} 0 & \text{for } -\infty < x < x_0 \\ 1 & \text{for } x \geq x_0 \end{cases}, \tag{B.10}$$

the output function, $g(x)$, will be the convolution of $f_i(x_0)$ with the LSF, $L(x - x_0)$:

$$g(x) = N\int_{-\infty}^{+\infty} f_i(x_0)L(x - x_0)\mathrm{d}x_0 = f_i(x_0) * L(x). \tag{B.11}$$

From the "convolution theorem" (see Appendix A, equation (A.34) applied to one-dimensional functions, one can obtain:

$$f_i(x) * L(x) \Leftrightarrow F(u) \cdot H(u) \tag{B.12}$$

where function $H(u)$ is the Optical Transfer Function (OTF) of the system, as will be explained later in Section B.4. Its modulus, $|H(u)|$, is defined as the Modulation Transfer Function (MTF) of the system.

*In this case (one-dimensional geometry), the MTF is the modulus of the Fourier Transform of the Line Spread Function.*

## B.3. Measurement of the Modulation Transfer Function for a linear system

There are several ways of measuring the MTF of a system. Here we give some indications of how to proceed for the MTF of a digital camera.

The starting point consists in the issues explained above, *i.e.* the modulus of the Fourier Transform of the LSF of a system is the MTF. In our case, the LSF can be obtained by differentiating the ESF. The main problem is therefore measuring the ESF as accurately as possible.

One can use the following procedures.

- Keep an image of an edge by the camera. Be sure that the edge is very sharp and the material is almost black to the radiation used (An edge that is good for low energy X-rays is not usually a good edge for X-rays produced by a LINAC).[10]
- Extract numerical data corresponding to a line crossing the edge perpendicularly.
- Correct for the noise due to background and CCD defects.
- Differentiate the obtained curve to get the LSF. (Some researchers prefer to fit the obtained LSF with a Gaussian, then to perform the FFT of the Gaussian obtained by the fit).
- Calculate the modulus of the Fast Fourier Transform of the LSF to obtain the MTF.

It has to be remembered that what we measure is the MTF of the system. In this case: the defects of the CCD sensor, the lens aberration, the diffused component of the radiation, the "penumbra" caused by the non-punctiform focal spot of the X-ray tube, and so on, are combined.

However, the calculation of the MTF function from experimental data requires a deeper knowledge of the discrete implementation of the Fourier Transform and its output. This is not dealt with herein; for a detailed description of the discrete Fourier transform and the calculation of the MTF by means of the edge technique, see works by Kak and Slaney [32] and Fujita *et al.* [48].

## B.4. Modulation Transfer Function: general definition

Let us assume that we want to reproduce an object, which can be described as a function $I(x_0, y_0)$ in the object plane, represented by the subindex "0". What we will record is another function, let us say $G(x_i, y_i)$, in the image plane represented by the subindex "$i$". In general, we can write:

$$G(x_i, y_i) = \int\int_{-\infty}^{+\infty} H(x_i, y_i; x_0, y_0) I(x_0, y_0) \mathrm{d}x_0 \mathrm{d}y_0 \tag{B.13}$$

The function $H(x_i, y_i; x_0, y_0)$ summarises all the defects and aberrations of our detection equipment, responsible for the imperfect reproduction of the image of the object. Usually, the acquisition process is supposed to be linear and invariant for translation, so that $H$ has the form:

$$H = H\,(x_i - x_0, y_i - y_0). \tag{B.14}$$

[10] For X-rays generated by a LINAC, tungsten or tantalum is usually used (at least 10 cm thick).

In this case, equation (B.13) assumes the form:

$$G(x_i, y_i) = \int_{-\infty}^{+\infty}\!\!\int H(x_i - x_0, y_i - y_0) I(x_0, y_0) \mathrm{d}x_0 \mathrm{d}y_0 \tag{B.15}$$

The integral in the right hand side of equation (B.15) is the convolution of the function $I(x_0, y_0)$ with the function $H(x_i - x_0, y_i - y_0)$, so that one can formally write:

$$G = I * H \tag{B.16}$$

Now consider the normalised Fourier transform of either $I(x_0, y_0)$ or $G(x_i, y_i)$ defined as:

$$I(\omega_x, \omega_y) = \frac{\int_{-\infty}^{+\infty}\!\!\int I(x_0, y_0) \exp\{-i(\omega_x x_0 + \omega_y y_0)\} \mathrm{d}x_0 \mathrm{d}y_0}{\int_{-\infty}^{+\infty}\!\!\int I(x_0, y_0) \mathrm{d}x_0 \mathrm{d}y_0} \tag{B.17}$$

$$G(\omega_x, \omega_y) = \frac{\int_{-\infty}^{+\infty}\!\!\int G(x_i, y_i) \exp\{-i(\omega_x x_i + \omega_y y_i)\} \mathrm{d}x_i \mathrm{d}y_i}{\int_{-\infty}^{+\infty}\!\!\int G(x_i, y_i) \mathrm{d}x_i \mathrm{d}y_i} \tag{B.18}$$

Remembering the convolution theorem, one gets:

$$G(\omega_x, \omega_y) = I(\omega_x, \omega_y) \cdot H(\omega_x, \omega_y), \tag{B.19}$$

where

$$H(\omega_x, \omega_y) = \frac{\int_{-\infty}^{+\infty}\!\!\int H(x, y) \exp\{-i(\omega_x x + \omega_y y)\} \mathrm{d}x \mathrm{d}y}{\int_{-\infty}^{+\infty}\!\!\int H(x, y) \mathrm{d}x \mathrm{d}y}. \tag{B.20}$$

$H$ is known as the Optical Transfer Function (OTF) and is a complex function.

The absolute value of the OTF, $|H|$, is called as the Modulation Transfer Function (MTF) of the optical system. As the Fourier Transform of a Gaussian function is also a

Gaussian function, the MTF also usually has a Gaussian shape. The variables of the MTF are the spatial frequencies (usually in lp/mm) along two perpendicular directions and are directly related with the spatial resolution of the system. In practice, MTF gives fraction of the amplitude of a modulated signal, ranging from 100 to 0%, as a function of spatial frequency. Obviously, higher the frequency the worse the signal reproduction is. On the Internet, one can find instructive examples of MTF. For instance, in Ref. [48], it is possible to see the representation of MTF when a "bar pattern" and a "sine pattern" are reproduced by an optical system (film + lens).

## APPENDIX C: CHARACTERISTICS OF SOME DETECTION SYSTEMS

### C.1. General considerations

Many systems are used in digital imaging. In the text, we stated that there are two commonly used categories: "flat panels" and CCD + lens + scintillator systems. We will give an overview of the flat panels currently (mid-2004) available on the market and a guideline to choosing CCD system components suitable for diagnostic applications in cultural heritage. As this field is growing fast, we do not aim to provide complete information but merely give suggestions of how to tackle the question and how to plan searches on the Internet.

### C.2. Flat panels

Flat panel digital detectors were introduced in the late 1990s as an alternative technology to traditional film, computed radiography and image intensifiers for medical imaging. These devices incorporate an X-ray detector and an integral sensor in a relatively compact design, hence the name flat panel. The detector is either an X-ray photoconductor or a scintillator, and the sensor is an amorphous silicon (a-Si) thin film transistor (TFT) array. Commercially available detectors use CsI or GOS scintillators coupled to an amorphous silicon TFT or to a CMOS photodiode array. Flat panels can basically be divided onto two types: direct and indirect. Both types are based on thin layers of a-Si deposited onto the glass substrates, with arrays of detector elements fabricated on the a-Si. For indirect conversion, the X-rays are first converted into visible photons, and then these photons are converted into electric charge. The scintillator, which converts X-rays to visible photons, is either grown directly on or attached to the TFT panel. For direct conversion flat panels, a different approach has been chosen: here the X-rays are directly converted into electric charges, which are then collected by the TFT array. Materials for direct conversion flat panels include amorphous selenium (a-Se), cadmium telluride (CdTe) and mercuric iodide ($HgI_2$).

As the performance of flat panels has improved and their cost has decreased, they have begun to challenge conventional imaging techniques, both in medical imaging and in other areas, such as industrial inspection and neutron imaging. These systems potentially offer a number of advantages over existing detector technologies, such as very compact size, large sensitive areas, and improved image quality under a wide range of imaging conditions. Flat panels offer substantial benefits for many X-ray imaging applications, in terms of resolution

**Table C.1.** Characteristics of some flat panels of a-Si + scintillator

| Manufacturer – Model | Varian – PAXSCAN 4030R | General Electric – RADView Si40 | Thales – FlashScan | PerkinElmer – RID1640 |
|---|---|---|---|---|
| Total area | 28 × 40 cm | 28 × 40 cm | 29 × 40 cm | 40 × 40 cm |
| Scintillator screen | GOS | GOS | GOS | GOS (standard) or CsI (optional) |
| Pixel number | 2304 × 3200 | 2304 × 3200 | 2240 × 3200 | 1024 × 1024 / 2048 × 2048 |
| Pixel size | 127 μm | 127 μm | 127 μm | 400–200 μm |
| Output | 12 bit | 12 bit | 14 bit | 16 bit |
| Time/frame | ~5 s | 3.4 s | 1.4 s | 0.3 s |

and dynamic range. In addition, they are physically robust and have good performance in terms of low noise and short readout time.

Commercial flat panels available on the market have a sensitive area up to 40 × 40 cm, with pixel size in the range between 50 and 400 μm. The output signal is usually 12 or 16 bit, and the number of pixels is comprised between $1 \times 10^6$ and $6 \times 10^6$. The main applications are in medical imaging, where the voltage used is below 150 kV. However, flat panels could also be used, with proper adjustment, in high energy X-ray imaging and neutron imaging. Table C.1 contains the characteristics of some a-Si flat panels suitable for medium-high energy X-ray imaging.

## C.3. CCD-based systems

The detection system described herein was developed by the Department of Physics of the University of Bologna for an important cultural heritage conservation institute. It is used mainly for diagnostics applied to ancient objects of archaeological interest and consists of several separate elements (see Fig. C.1):

- a metallic box with an internal guide for the movement of the CCD camera appropriately shielded by lead;
- a scintillating screen on which the X-ray beam generates the image;
- a mirror, angled at 45°, which reflects the image towards the camera;
- the CCD camera;
- a collimator, located in front of the screen, which decreases the radiation diffused by the object (sometimes the most important cause of image degradation);
- a pre-collimator, placed close to the X-ray source, which moulds the beam.

Each component must be chosen bearing in mind the energy range of the X-rays. DR and CT of objects of interest in the cultural heritage field are usually performed using

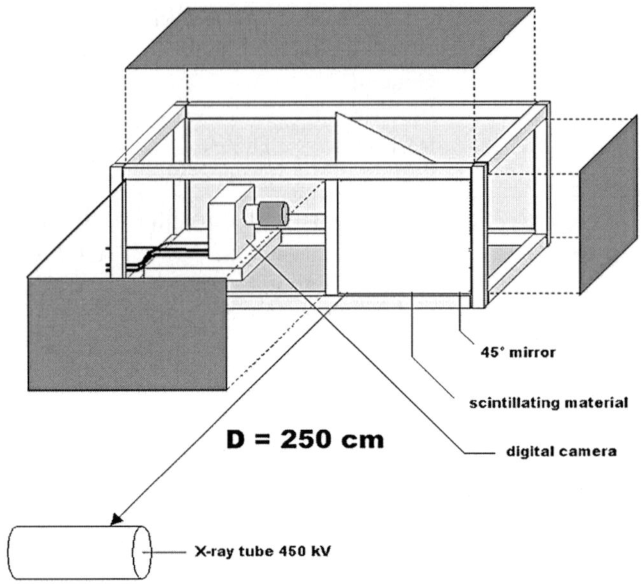


**Fig. C.1.** Detection system composed by scintillator screen + mirror + lens + CCD camera. Lead sheets protect the camera from scattered photons.

450 kV X-ray tubes or by 9 MeV LINAC. For X-ray tubes, the Pb should be at least 30 mm thick for both collimators and camera shielding. When using LINAC, the Pb should be at least 100 mm thick. Moreover, a sheet of lead glass must be placed in front of the camera to shield the photons that could reach the CCD sensor through the lens.

*Scintillating screen*

This is an extremely important component and should have the following properties:

- high light output per unit of energy dissipated within it (light photons released per MeV);
- low afterglow;
- high stability (chemical, temperature, hygroscopicity and radiation damage);
- high absorbing coefficient for X-ray photons;
- high MTF (good spatial resolution);
- emission wavelength well-matched to CCD sensitivity curve;
- not excessively expensive when used on large areas (*e.g.* $30 \times 40$ cm$^2$).

A new material that satisfies all these requirements is caesium iodide (CsI) that is "structured" to form needles (Hamamatsu).

Another good scintillator is gadolinium oxysulfide (named GOS), which is cheaper than CsI. It can be smeared on a heavy metal sheet; so that light can also be produced by the

**Table C.2.** Principal characteristics of some scintillator materials

| Scintillator | $\rho$ [g/cm$^3$] | Refractive index $n$ | Peak emission [nm] | Decay time [µs] | Light output [ph/MeV] | Radiation hardness [Gy] | Hygroscopic/ mechanical treatment |
|---|---|---|---|---|---|---|---|
| CsI (Na) | 4.51 | 1.84 | 420 | 0.63 | 38500 | $<10^6$ | yes/good |
| CsI (Tl) | 4.51 | 1.80 | 550 | 0.9 | 59000 | $<10^6$ | slightly/good |
| NaI (Tl) | 3.67 | 1.85 | 415 | 0.23 | 38000 | $10^2$ | yes/good |
| BGO ($Bi_4Ge_3O_{12}$) | 7.13 | 2.15 | 480 | 0.3 | 8000 | $10^5$ | no/very good |
| GSO ($Gd_2SiO_5$) | 6.71 | 1.9 | 440 | 0.06 | 8000 | $>10^8$ | no/bad |
| GOS ($Gd_2O_2S$) | 7.34 | 2.2 | 510 | 3 | | | no/good |
| CWO ($CdWO_4$) | 7.9 | 2.25 | 470 | 20 | 14000 | $10^6$ | no/bad |
| LSO ($Lu_2SiO_5$) | 7.4 | 1.82 | 420 | 0.04 | 25000 | $>10^7$ | no/acceptable |
| GS1 glass | 2.64 | 1.58 | 395 | 0.055 | 4000 | $10^5$ | no/good |
| TB2 glass | 2.64 | 1.58 | 550 | 3.5 | 6000 | $\approx 10^6$ | no/good |
| $(Y,Gd)_2O_3$:Eu | 5.9 | | 611 | 1000 | 19000 | | |
| $Gd_2O_2S$:Pr | 7.3 | | 513 | 3 | 21000 | | |

photo-electrons extracted from the metal layer, which dissipate their energy in the GOS. The LSF of CsI in needles is sharper than that of GOS.

Fiber-optic scintillator (FOS) are often used to obtain high spatial resolution. This scintillator is made with a plate of scintillating optical fibres sometimes covered by Lanex (another GOS-based material) (Kodak). The core of the scintillating fibre is made of heavy terbium activated glass.

In recent times, new matrix scintillators have been put in the market. These include the LSO (Lutetium oxyorthosilicate ($Lu_2SiO_5$:Ce)) and BGO (Bismuth Germanate ($Bi_4Ge_3O_{12}$)) models. Previously, these two scintillators were commonly used as single crystals but not as a matrix. The main characteristics of some scintillating materials are reported in Table C.2.

*CCD camera*

The camera, coupled with a lens, focuses the light emitted from the scintillating screen on the CCD chip and captures the image. The light emitted from the screen is reflected to the camera by a mirror. The mirror used must have high reflectivity (>95%) for the wavelength of the light emitted by the scintillator screen and no residual activation. Moreover, the wavelength of the light emitted by the screen must be in the range where the CCD has greatest sensitivity, usually around 550–600 nm (see Fig. C.2).

The digital camera is connected to a computer to upload the captured images stored in its memory. The PC interface could be a digital frame grabber or a standard PC connection (USB or Ethernet).

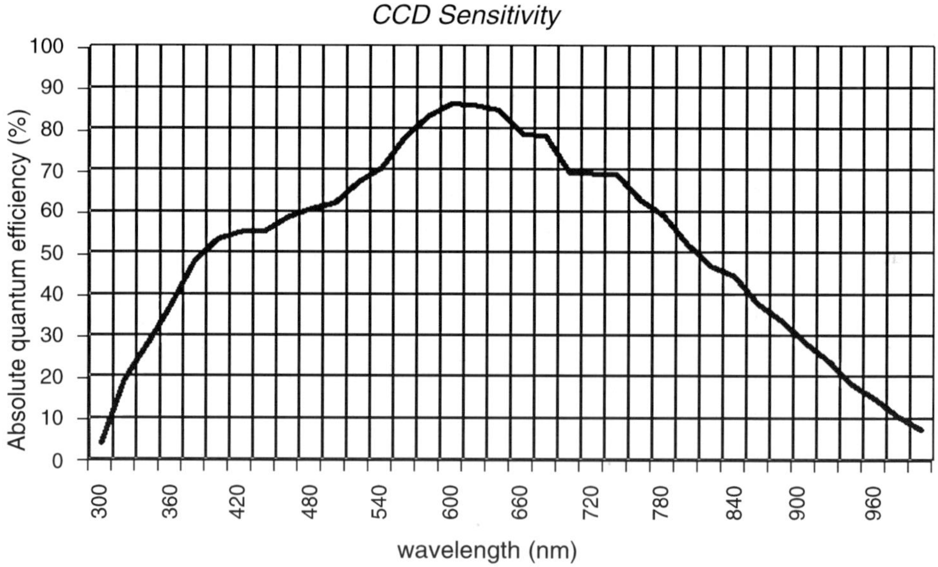


**Fig. C.2.** Efficiency of a CCD as a function of the wavelength (courtesy Apogee Instruments INC)

A CCD camera for digital imaging applications must have low noise and high efficiency. Depending on the budget available and on the parameters of the CT scans to be performed, one may choose one of several CCD sensors with different numbers of pixels, bits, pixel-sizes, quantum efficiencies, wavelength sensitivities and so on. As tomographic analysis requires a large number of good quality images (up to 1000), the read-out time for each frame is a very important parameter.

If only a little light is produced in the scintillator or if one wishes to acquire the images in a short time, an intensified camera (*i.e.* the Electron Bombarded CCD) can be used. Conversely, this kind of detector is affected by a greater electronic noise. In order to decrease the noise level or, better still, to suppress the dark current, CCD cameras have excellent cooling systems that can keep chip temperature at about 50°C lower than room temperature (−130°C with nitrogen).

In most cases, CCD cameras need to be coupled with a lens to focus the image on the chip and therefore the high brightness of the lens is very important in decreasing the exposure time. The larger the aperture of the lens, the greater is the difficulty in focusing the camera. In such cases, a micrometric stage is required. The best approach is to perform this operation directly from the control room of the bunker, on the radiographic image produced by X-rays on the scintillating screen. The picture distortion due to the lens (barrel or pin-cushion) can be corrected using software.

Otherwise, a CCD camera can acquire images without the lens by being coupled directly with a fibre optic plate or taper. To avoid image resolution loss, bond joints on CCDs must be perfectly matched. Moreover, in order to improve resolution and decrease cross-talk, one may choose from different kinds of fibre arranged in different configurations. Fibre optic bundles with individual fibre diameters as small as 3 microns can be bonded to CCD.

## ACKNOWLEDGEMENTS

Much of the work described in this chapter was initiated by Massimo Rossi, a young researcher who tragically died from a terrible illness a few years ago, at the age of 35. This work is dedicated to him. The author wishes to thank all his young contributors, pictured in Fig. 69 (From right to left, they are: (stand up) Alessandro Pasini, Nico Lanconelli, Matteo Bettuzzi, Samantha Cornacchia, Maria Pia Morigi, Marilisa Giordano, Alice Miceli, the author; (sit down) Alessandro Fabbri, Davide Bianconi, Carlotta Cucchi, Emilia di Nicola. Not in picture: Davide Romani, Alberto Rossi, Rossella Brancaccio), for their enthusiasm in applying advanced techniques in the interest of the conservation and understanding important cultural treasures. Heartfelt thanks to Serena Pini (Musei Comunali, Florence) and to Giacomo Chiari, Senior Scientist of the Getty Conservation Institute, for their much appreciated collaboration.

## REFERENCES

[1] Y.V. Kuzumin, V.K. Popov, M.D. Glascock, M.S., Shackley, Sources of archaeological volcanic glass in the primorye (Maritime) province, Russian Far Eastl, *Archaeometry*, **44** (2002), 505–515.

[2] C.C. Graham, M.D. Glascock, J.R. Vogt, T.G. Spalding, Determination of elements in the National Bureau of Standards' geological standard reference materials by neutron activation analysis, *Analytical Chemistry*, **54**, (1982), 1623–1627.

[3] W. Kockelman, Principles and Applications of Time-of-Flight Neutron Diffraction (TOF-ND) for Characterising Archaeological Ceramics and metal Artefacts. http://www.ifac.cnr.it/corsoArcheo/abstract/Kockelmann.pdf
[4] *Radiation Protection Design Guidelines for 0.1–100 MeV Particle Accelerator Facilities*, NCRP Report No. 51.
[5] VARIAN-1, http://www.varian.com/sein/lnm000.html
[6] Sincrotrone ELETTRA, Trieste, Italy, http://www.elettra.trieste.it/info/index.html
[7] European Synchrotron Radiation Facility (ESRF), Seeing Things in a Different Light, http://www.esrf.fr/info.
[8] M. Pantos, Synchrotron radiation in archeological and cultural heritage sciences, in: *4th Symposium on Archaeometry*, Athens, 2003.
[9] S. Bracci, F. Falletti, M. Matteini, R. Scopigno, *Exploring David*, Ediz. Giunti Firenze Musei, pp. 221–226, 2004.
[10] B. Illerhaus, J. Goebbels, H. Riesemeier, *3D Computerized Tomography – Synergism Between Technique and Art*, http://www.ct.bam.de/daten/paper.html
[11] H.E. Martz Jr., C.M. Logan, P.J. Shull, Radiology, in: P.J. Shull (Ed.), *Nondestructive Evaluation – Theory, Techniques, and Applications* (ISBN: 0824788729), The Pennsylvania State University, 2002.
[12] NDT Resource Center, http://www.ndt-ed.org/index_flash.htm
[13] R.E. Woods, R.C. Gonzalez, Real-Time Digital Image Enhancement, in: *Proceedings of the IEEE*, **69** (1981) 5, 643–657.
[14] S. Di Zenzo, Advances in Image Segmentation, *Image and Vision Computing*, **1** (1983) 2, 196–210.
[15] M. Rossi, F. Casali, P. Chirco, M.P. Morigi, E. Nava, E. Querzola, M. Zanarini, X-ray 3D computed tomography of bronze archaeological samples, *IEEE Transactions on Nuclear Science*, **46** (1999), 897–903.
[16] H.C. Andrews, B.R. Hunt, *Digital Image Restoration*, Prentice-Hall, Englewood Cliff, New Jersey, 1977.
[17] R.C. Gonzalez, P. Wintz, *Digital Image Processing*, Addison Wesley, Massachusetts, 1977.
[18] R.C. Gonzalez, R.E. Woods, *Digital Image Processing*, second Ed., Prentice Hall, New Jersey, 2002.
[19] W.K. Pratt (Ed.), *Digital Image Processing*, Strandberg, Denmark, 1985.
[20] A. Rosenfeld, A.C. Kak, *Digital Picture Processing*, Academic Press, New York, 1976.
[21] E.O. Brigham, *The Fast Fourier Transform*, Prentice-Hall, Englewood Cliffs, New Jersey, 1988.
[22] V. Cappellini, *Digital filters and their applications*, Academic Press, 1981.
[23] M. Rossi, F. Casali, S.V. Golovkin, V.N. Govorun, Digital radiography using an EBCCD-based imaging device, *Applied Radiation and Isotopes*, **53** (2000), 699–709.
[24] Varian Paxscan 2020, http://www.varian.com/xray/pdf/paxscan2020.pdf
[25] AETNA, http://www.aetna.com/cpb/data/CPBA0380.html
[26] M. Bettuzzi, A. Pasini, A. Rossi, D. Schneberk, private communication, October 2004.
[27] P. Von der Hardt, H. Rottger (Ed.), Neutron Radiography Handbook, D. Reidel Publishing Company, 1981. Revised by E. Heiberg, Neutron Radiography. International Advances in NDT. **Vol. 16**, 1991.
[28] Applied Science, www.appscientech.com
[29] R. Massari, R. Rosa, Neutron Tomography at the TRIGA reactor of ENEA, http://info.casaccia.enea.it/triga/TRITON/.
[30] Doyle Group Harvard, http://www.doylegroup.harvard.edu/neutron/publications/Theses/Park.pdf
[31] Imatron CT, http://www.vaytek.com/heart1.htm
[32] A.C. Kak, M. Slaney, *Principles of Computerized Tomographic Imaging*, The Institute of Electric and Electronics Engineers, Inc., New York, NY, IEEE Press, 1988, Sec 6.4.1.
[33] M. Bettuzzi, F. Casali, S. Cornacchia, M. Rossi, E. Paltrinieri, M.P. Morigi, R. Brancaccio, D. Romani, A new linear array detector for high resolution and low dose digital radiography, in: Proceedings IRRMA-V conference, Bologna, June 9–14, 2002, *Nuclear Instruments and Methods in Physics Research* B, **213** (2004), 227–230.
[34] ARACOR, http://www.aracor.com/pages/rocketmotor.html
[35] Bio-Imaging, http://www.bio-imaging.com/detectors.asp
[36] General Electric High Speed CT, http://www.bmimed.com/medical-imaging-products.asp
[37] L.A. Feldkamp, L.C. Davis, J.W. Kress, Practical cone-beam algorithm, *Journal of the Optical Society of America*, **1(A)** (1984), 612–619.
[38] M. Rossi, F. Casali, M. Bettuzzi, M.P. Morigi, D. Romani, S.V. Golovkin, V.N. Govorun, An experimental micro-CT system for X-ray NDT, *SPIE Proceedings*, **4503** (2002), 338.
[39] M. Rossi, F. Casali, D. Romani, L. Bondioli, R. Macchiarelli, L. Rook, MicroCT Scan in paleobiology: application to the study of dental tissues, *Nuclear Instruments and Methods in Physics Research* B, **213** (2004), 747–750.

[40] A. Pasini, F. Baruffaldi, M. Bettuzzi, R. Brancaccio, F. Casali, S. Cornacchia, N. Lanconelli, M.P. Morigi, E. Di Nicola, S. Pani, E. Perilli, D. Romani, A. Rossi, A CCD-based high resolution CT system for analysis of trabecular bone tissue, in: *IEEE Medical Imaging Conference*, Rome, October 16–22, 2004.

[41] Hitachi Heavy Industry, *Industrial X-ray CT for Digital Engeneering*, http://www.hitachi.com.

[42] Paul Scherrer Institute, Neutron tomography http://neutra.web.psi.ch/What/tomo.html

[43] T. Nakano, Y. Kawabata, M. Hino, U. Matsushima, Development of CT imaging system with very low energy neutrons, http://www.ttp.net/abstract_OR_prevvauth.htm, 2004.

[44] El-Bakkoush (2000), http://www.ndt.net/article/wcndt00/papers/idn704/idn704.htm

[45] Filippo Giovanelli, *Tecniche di analisi non distruttive nello studio e nella tutela del patrimonio artistico*, Thesis, Università degli studi di Roma "La Sapienza", Dipartimento di Ingegneria Nucleare e conversioni di energia, 2000/2001.

[46] E. Deschler-Erb, E.H. Lehmann, L. Pernet, P. Vontobel, S. Hartmann, The complementary use of neutrons and X-rays for non-destructive investigation of archaeological objects from Swiss collections, *Archaeometry*, **46** (2004) 4, 647.

[47] A.A. Harms, D.R. Wyman, *Mathematics and Physics of Neutron Radiography* (ISBN: 9027721912), Kluwer Academic Publishers, The Netherlands, 1986.

[48] H. Fujita et al., A simple method for determining the Modulation Transfer Function in digital radiography, *IEEE Transactions on Medical Imaging*, **11**, No. 1, March 2000.

[49] N. Koren, http://www.normankoren.com/Tutorials/MTF.html

[50] J.H. Hubbell, S.M. Seltzer, Tables of mass attenuation coefficients for X-rays, http://physics.nist.gov/PhysRefData/XrayMassCoef/cover.html

## Chapter 3

# Investigation of Diagenetic and Postmortem Bone Mineral Change by Small-Angle X-Ray Scattering

Jennifer C. Hiller and Tim J. Wess

*Structural Biophysics Group, School of Optometry and Vision Science, Cardiff University, King Edward VII Avenue, Cardiff, CF10 3NB, UK*
*Email: j.c.hiller@stir.ac.uk*
*Email: WessTJ@cardiff.ac.uk*

**Abstract**

Alteration to bone in the burial environment is inevitable, and affects many aspects of archaeological scientific investigation. The mineral component of bone contributes to its postmortem longevity, and may be a resource for study in its own right. Here we review recent work in our laboratory, in which we have established the use of small-angle X-ray scattering (SAXS) to detect alteration to bone crystallites in the postmortem and depositional environments. We have had success linking diagenetic crystallite change to the preservation of ancient biomolecules. Moreover, we have developed microfocus mapping techniques to track alteration in bone thin sections across microscale areas such as individual osteons. Finally, we describe our use of SAXS to measure alteration to bone mineral in experimental heating regimes designed to mimic burning or cremation. These results are an indication of the potential for SAXS to elucidate postmortem and diagenetic alteration to bone mineral, which is a valuable resource for archaeological science.

**Keywords:** Bone, X-ray scattering, ancient biomolecules, archaeological science.

**Contents**

# 1. INTRODUCTION AND CONTEXT

## 1.1. Bone diagenesis and biomolecular preservation

Bone is a composite material, consisting of a mineral phase (primarily carbonated hydroxyapatite) embedded in and overgrowing an organic (mostly proteinaceous) fraction. The primary organic component of bone is type I collagen, composed of three helical polypeptide chains wound together into a triple helical structure. The triple helices are arranged in bundles (fibrils) with a staggered spacing, and grouped together to form fibres of a tightly woven rope-like structure. Bone mineral primarily consists of a carbonated form of hydroxyapatite, $Ca_{10}(PO_4)_6OH_2$ [1]. The bioapatite crystallites are non-stoichiometric and very small, between 2 and 4 nm thick in the smallest dimension in modern bone [2]. The mechanism of biomineralisation in bone is still not understood, but results from modelling [3] as well as from similar systems such as mineralising turkey leg tendon [4,5] indicate that the mineral crystallites form initially in the hole or "gap" regions in the collagen fibrils. Once nucleated, the crystals elongate along the collagen fibril, remaining primarily oriented along the long axis of the bone, although this can vary slightly [6]. The mineralised fibres are then packed into lamellae [7], which in the osteonal bone are then laid down in a "rotated plywood" architecture of concentric rings surrounding the central Haversian canal [8].

Bone, therefore, obeys a strict hierarchical organisation from the nanoscale to the macroscale. Furthermore, due to lattice imperfections and instability in the crystallites, a certain thermodynamic instability is inherent in the structure. It is the composite nature of bone and the associated mineral structure that lends it more robusticity than observed in soft tissue following burial, but degradation still occurs. Diagenesis can begin with either the mineral fraction (dissolution or demineralisation) or the collagen (microbial colonisation or chemical degradation) [9], but either way, the initial alteration to the structure may initiate a cycle of mineral and protein loss that ends in the complete degradation of the bone.

Studies of archaeological bone diagenesis have focused on both the mineral and organic fractions of bone tissue in an attempt to elucidate the mechanisms of preservation of archaeological biomolecules [9]. Microbial damage in bone has been seen to cause foci of severe mineral alteration, while neighbouring bone remains intact [10,11]. However, it is not known how much variation takes place across a region of bone tissue, and most diagenetic analyses are conducted on powdered samples or fragments. While methods such as microfocus infrared spectroscopy have been successfully employed to examine variation in fresh modern bone [12,13], the application of these techniques to archaeological samples has often been limited by difficulties in sample preparation.

Alteration to bone mineral can present a significant danger to the preservation of biomolecules in bone, particularly if the molecules in question are bound to or protected by the mineral matrix. Due to the importance of bone mineral in the preservation of biomolecular information over archaeological time, screening techniques that examine the diagenetic state of the mineral fraction have been developed. These include the use of Fourier-transform infrared spectra (FTIR) to calculate indices of crystal change [14,15]; X-ray diffraction (XRD) to determine crystal composition and strain as well as bone density [16,17]; and particle-induced X-ray emission (PIXE) to examine diagenetic chemical substitutions in the mineral structure [18], among others. Some have proposed the use of indices of crystal alteration for relative dating techniques [19]. A correlation between the FTIR splitting factor (defined in Ref. [20]) and the level of organic preservation represented by weight percent nitrogen has been shown [21], reinforcing the view that alteration in the mineral component of bone contributes to the loss of the organic component. Moreover, it has been suggested that DNA preservation in archaeological remains could also be subject to change in the mineral component [22]. However, there is evidence to suggest that the results of some of the techniques employed to measure mineral alteration can be partially dependent on sample preparation procedures [23].

Small-angle X-ray scattering (SAXS) has recently found a new application in the study of mineral change in archaeological bone. This allows for the accurate determination of crystal size, shape and orientation within bone and is independent of crystal lattice perfection [2,24,25]. SAXS has been shown to provide information regarding crystallite structure that is complementary to other techniques such as FTIR [26], which are employed in archaeological contexts, and recently has been used to characterise diagenetic change in bone and other materials of archaeological interest [11,27,28]. The measurement of subtle changes to crystallite thickness or shape can provide a window to the events leading to mineral alteration in archaeological bone, including diagenetic processes as well as human interventions such as burning [29] and some types of mummification [30].

### 1.2. Introduction to small-angle X-ray scattering

A full treatment of the physical basis behind SAXS methods is outside the scope of this discussion, but is presented comprehensively in several texts, including Refs. [31] and [32]. A brief introduction to the theory behind SAXS will be presented here, with the primary aim of explaining the measurements used to determine the alteration of bone mineral in archaeological samples. The method is considered to be useful for the examination of bone crystallites due to their size distribution. SAXS is also not influenced by lattice perfection, and produces an average measurement over a volume of bone [2].

X-rays are scattered by all matter. X-ray diffraction occurs when a beam impinges on a regular and highly ordered atomic lattice at a relatively high angle of incidence. At lower angles of incidence, scattering occurs from electron density contrast between the particle and the surrounding medium. The correlation between larger particles, on the nanoscopic scale, tends to be more irregular than interactions on the atomic scale. Consequently, the sharp diffraction peaks observed in high-angle X-ray diffraction range, which result from ordering within a crystallite, are complemented by a more diffuse scattering in the small-angle X-ray scattering range. At this scale, it is the electron density contrast between interfaces on the nanoscale that scatters the main incident beam. This scattering effect can be used to determine shape, size, orientation and packing of objects on the nanoscale. The small-angle X-ray scattering examination of bone is enabled by the electron density contrast between the mineral and collagenous components [33].

X-ray diffraction from a crystalline lattice follows the well-known Bragg's Law:

$$2d \sin\theta = n\lambda \tag{1}$$

where $d$ is the interplanar distance in the crystal, $\theta$ the angle of incidence and reflection (Bragg angle) and $\lambda$ the wavelength of the incident X-ray [34]. Typically, X-ray diffraction results are given as arbitrary units of intensity plotted against $2\theta$; crystal lattice reflections are

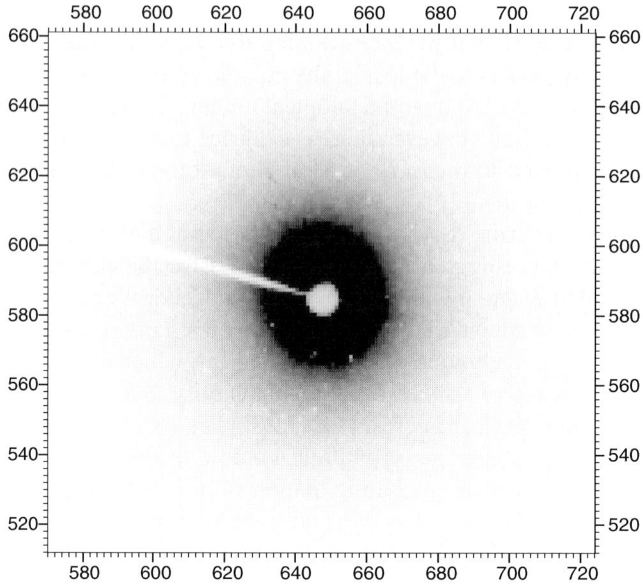


**Fig. 1.** Example of an X-ray scattering image taken from a thin section of fresh bovine bone, covering scattering angles from 0 to 6°$\theta$.

thus given in terms of the corresponding Bragg angle. As the sample becomes larger with respect to the wavelength of the incident light, the angle $\theta$ becomes smaller, until a small-angle X-ray scattering situation is reached for particles of colloidal dimensions (1–1000 nm), where the sample is much larger than the wavelength $\lambda$ [35]. Figure 1 is an example of a scattering image from fresh bovine bone taken at the European Synchrotron Radiation Facility (ESRF).

Theoretically, there is no maximum size for the particles that can be measured. For some structures, long range crystalline interactions result in diffraction peaks at increasingly small angles. However, in all matter, diffraction corresponding to the vectorless component of the object at the (000) reflection can be seen as a broadening of the main beam. Figure 2 illustrates this graphically, showing an incident beam and the scattering vectors produced.

Ideally, the incident beam will consist of highly parallel, monochromatic light to reduce distortion in the scattering results. This situation is approximated through collimation of the beam, followed by use of a monochromator, prior to the light reaching the sample. A two-dimensional detector is used to collect the scattering information, which allows for the examination of both isotropic (direction-independent) and anisotropic (direction-dependent) samples. Bone powders are typically isotropic, due to the loss of crystallite orientation in grinding, but thin sections can produce anisotropic scattering profiles.

Rather than working with measurements in $2\theta$, which is dependent on wavelength, the measurements used here have been expressed in terms of distance in reciprocal space, $q$, where

$$q = \frac{2\sin\theta}{\lambda} \times 2\pi \quad \text{or} \quad q = 2\pi d, \tag{2}$$

where $d$ is measured in reciprocal nanometers ($nm^{-1}$).

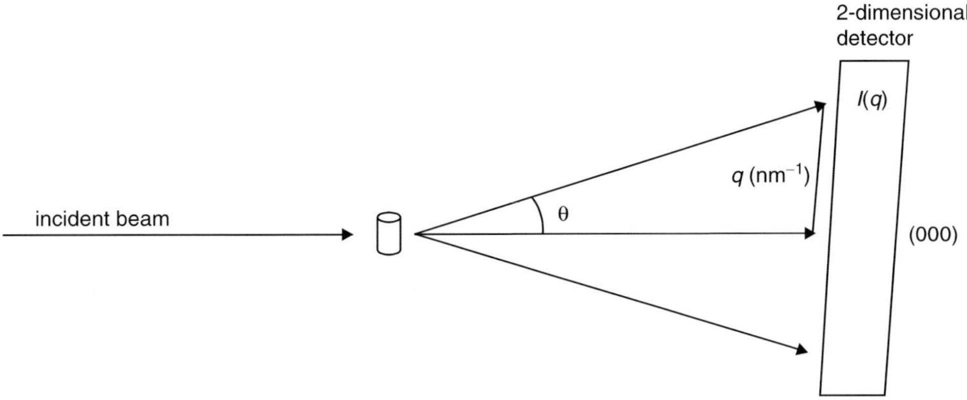


**Fig. 2.** Schematic diagram of scattering vectors.

For a given scattering result, a one-dimensional spherically averaged function of intensity $I(q)$ over distance in reciprocal space $q$ is generated. Even within the small-angle scattering region (typically angles up to 6°), the rate of change of intensity decay with respect to scattering angle varies due to three main factors: (1) the scatter resulting from the object itself (the form factor); (2) the interference between the scattering objects; and (3) the scattering resulting from the interfaces of objects. The scattering effects from such features can be revealed in a number of plotting routines. A plot of $I(q)\cdot q^4$ *versus* $q^4$ yields the background due to incoherent scattering as the slope and the Porod invariant $Q$ as the $y$ intercept [36]. If the particles present (in this case, bone crystallites) are taken to have a smooth, sharp electron density interface with the surrounding material, then the scattering from that surface will obey Porod's Law [37]. Here, the contribution of the scattering intensity that obeys a decay with the –4th power of the scattering angle, is an indication to the total area and perfection of interfaces in the sample, and can be used to calculate the surface to volume ratio of discrete particles [24]. The simplest derivation of Porod's Law of surface scattering is:

$$I(q) = \frac{S}{q^4} 2\pi Q \rho^2, \tag{3}$$

where $I(q)$ is the intensity of the scattering, $S$ the particle-specific surface, $Q$ the Porod invariant, and $\rho$ the difference in electron density between the two phases present (in this case, bone mineral and organic matrix) [37]. Figure 3 illustrates three different plots of $I(q)\cdot q^2$ *versus* $q$, with the start of the Porod region marked, where scattering intensity decays by the –4th power of the scattering vector $q$. For lower values of $q$, the shape of the curve is indicative of the shape of the scattering object such as cylinders or spheres, and is also modulated by the interaction between objects when the packing density is high. In the case of bone, we have assumed that the interaction between bone crystallites is low and the scattering curve can be used to determine crystallite shape, as discussed below.

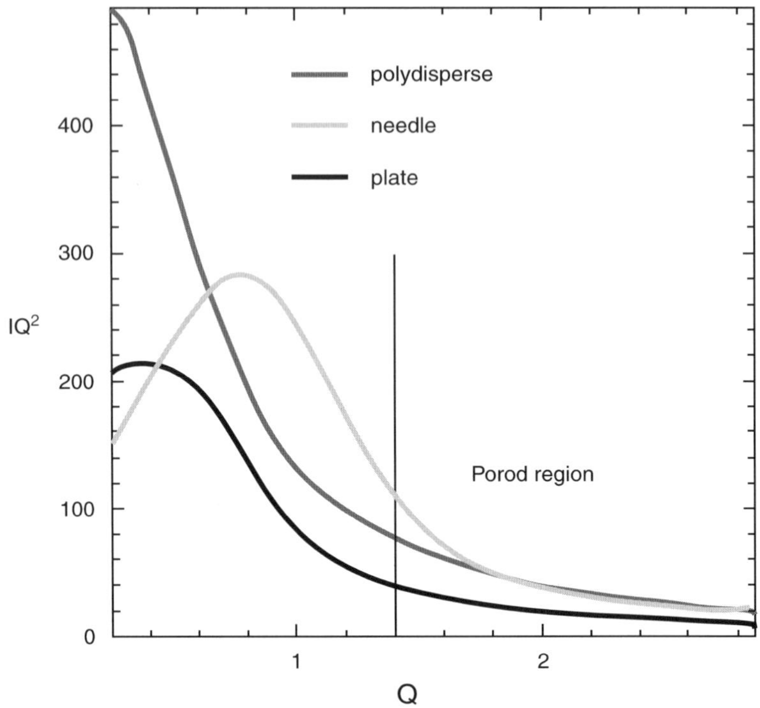


**Fig. 3.** An example plot (from Ref. [11]), showing the curve $I(q) \cdot q^2$ *versus* $q$, with the Porod region ($q^{-4}$) marked. For low values of $q$, the shape of the curve is used to determine crystallite shape (see below).

From the intensity of scattering, properties of crystallites in bone, including thickness, shape and orientation, can be calculated using techniques devised by Matsushima *et al.* [33,38] and Fratzl *et al.* [2,4,5,25]. For the analyses conducted here, crystallite thickness and shape will be considered. The orientation of bone crystallites could be determined from the scattering measurements made on bone thin sections, but was not systematically considered for the sake of this work. For powdered samples, orientation cannot be determined since the scattering profiles are isotropic.

### *1.2.1. Crystallite shape*

The curve of $I(q) \cdot q^2$ plotted against $q$ can be used to establish a shape parameter $\eta$, defined by Fratzl *et al.* [2] as the deviation of the realised curve from an ideal Lorentzian distribution. At small values of $q$, plate-like crystallites will exhibit a form factor curve that behaves as $q^{-2}$, whereas needle-like crystals exhibit a curve as $q^{-1}$ [4]. While $\eta$ itself is a continuum from polydisperse crystallites (little or no deviation from the Lorentzian) to needle- or rod-like crystallites (high deviation), arbitrary cut-off points for defining needle-like, plate-like, or polydisperse crystallite populations can be determined. Figure 3 illustrates the three types of curves derived from $I(q) \cdot q^2$, where the medium grey corresponds to polydisperse, black to

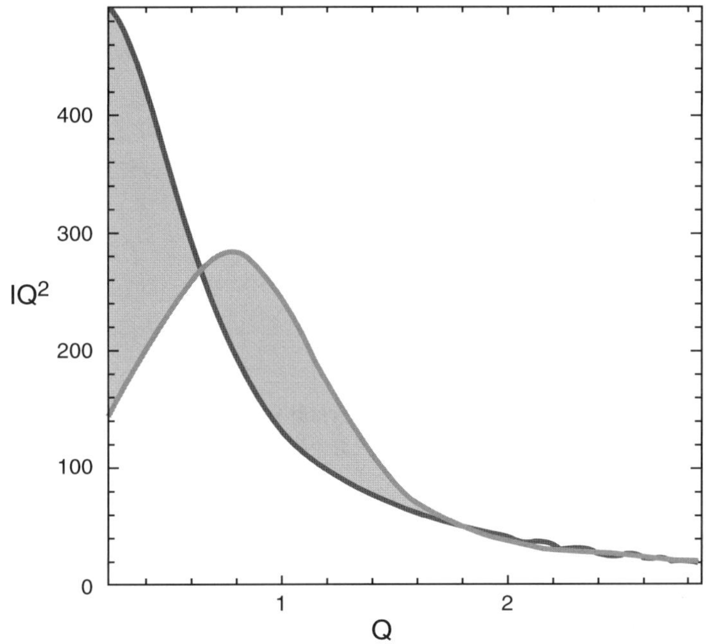


**Fig. 4.** $I(q) \cdot q^2$ versus $q$ plot altered to show the difference between the idealised Lorentzian function (black) and the curve for needle-like crystals (grey). The differences between the curves used to calculate $\eta$ are shaded; the squares of the areas are used.

plate-like and light-grey to needle-like crystallites. Figure 4 has been produced from the same figure, but the areas that are deviations between the Lorentzian distribution and a curve describing needle-like crystallites have been highlighted.

Starting from the ideal Lorentzian distribution, $\eta$ is calculated as the sum of the squares of the deviation seen in the calculated curve. This deviation is largest for needles, smaller for plates, and smallest (closest to the Lorentzian form) for polydisperse crystal populations. Figures 5 and 6 illustrate examples of curves for needle-like and plate-like crystallites, along with the $\eta$ values calculated for each curve.

The determination of a numerical parameter to characterise crystallite form allows for comparison between values without directly comparing curve shapes. This, in turn, has permitted the comparison of $\eta$ values with other diagenetic measurements, as well as the development of two-dimensional maps showing shape variation over an area of bone. The curves themselves were used to illustrate variation in shape along one-dimensional line scans across 200-μm long segments of bone, as described in Refs. [11,27]. Recent modelling, experiments have shown that, the form factor curves in bone can be representative of variations on a "stack of cards" motif of platelet crystallites, and the interpretation of needle-like crystallites may no longer be appropriate; this result could eventually force a reinterpretation of SAXS data in bone [39].

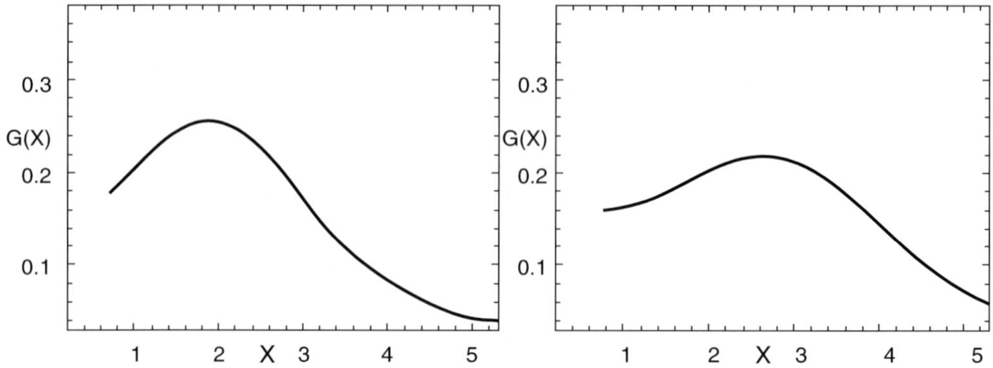


**Fig. 5.** Form factor curves for needle-like crystallites. For these curves, the calculated $\eta$ values were 0.0141 (left curve) and 0.0218 (right curve).

### *1.2.2. Crystallite thickness*

The thickness of bone crystallites, *T*, was defined as the thickness of the smallest dimension and was calculated as described in Refs. [2,4,5]. Crystal thickness is based on the integral of the curve $I(q)\cdot q^2$ as a function of $q$ (illustrated in Fig. 3), as well as the volume fraction of organic mineral present in the bone and the surface to volume ratio of the crystallite. For a crystallite with dimensions $a$, $b$ and $c$, then the thickness $T$ of the smallest dimension is characterised as

$$T = \frac{2ab}{(a+b)}, \quad \text{where } c >> a \text{ and } b \text{ (needle-like crystals),} \tag{4}$$

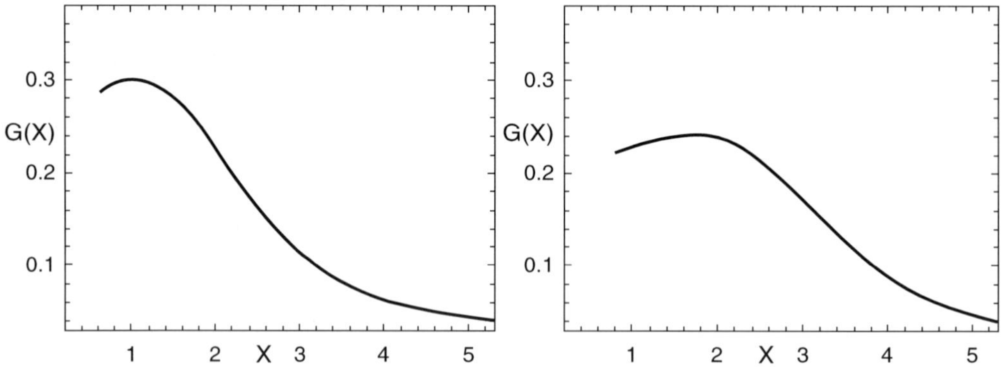


**Fig. 6.** Form factor curves for plate-like crystallites. For these curves, the calculated $\eta$ values were 0.00637 (left curve) and 0.0114 (right curve). The left curve, with its lower $\eta$ value and slightly different shape, is beginning to edge toward a polydisperse curve.

or as

$$T = 2a, \quad \text{where } a >> b \text{ and } c \text{ (plate-like crystals).} \tag{5}$$

Since the area under the curve $I(q) \cdot q^2$ is not dependent on the shape of the curve itself, the thickness of the smallest dimension can be calculated independent of the crystallite shape.

## 2. BIOMOLECULAR PRESERVATION

A set of archaeological bone samples collected as part of a larger study into the preservation of bone material in cave sites was used in this research. All bones were processed into finely ground powders by hand using an agate mortar and pestle. Three samples of modern cortical bone, two forensic human samples and one from a bear, were used as controls. For SAXS measurements, powdered samples (approximately 15 mg) were loaded into a sample carriage between two mica sheets and mounted in the vacuum chamber of the NanoSTAR (Bruker AXS, Karlsruhe) X-ray facility at Cardiff University. The data collection procedure used that followed described in detail in Ref. [27]. Scattering profiles were taken over 3 h exposures using a sample to detector distance of 1.25 m. Collected data were corrected for camera distortions, a background image was subtracted, and images were analysed using in-house software. The two-dimensional detector output was converted into spherically averaged one-dimensional profiles. Values for crystal thickness ($T$) of the smallest dimension, as well as curves describing crystallite morphology and a shape parameter ($\eta$), were determined for each sample using the SAXS data. A detailed procedure for these calculations is presented in Refs. [2,5].

Previous studies have established a link between various diagenetic parameters and the survival of ancient DNA, including histological preservation [40,41], amino acid racemisation [42], and protein preservation quantified with pyrolysis gas chromatography/mass spectrometry (Py-GC/MS) [43]. Moreover, the determination of the diagenetic condition of bone specimens has become a necessary hallmark of authentic ancient DNA results [44,45]. The role of stable hydroxyapatite surfaces in preservation of ancient DNA fragments has recently been considered [22] and, due to the tendency of apatite crystal surfaces to bind small polar molecules such as DNA fragments [46,47], preservation of intact bone mineral should contribute to the survival of ancient DNA. A crystal lattice and composition that resembles the biogenic structure as closely as possible seems to assist in DNA survival. Small-angle X-ray scattering (SAXS) is capable of measuring alteration to the real crystal structure, in a manner that is independent of lattice perfection or composition. The presence of a biogenic apatite structure and its relationship to ancient DNA preservation can thus be investigated.

Bone-crystallite thickness measurements were obtained from SAXS data. For these powdered samples, grinding had disrupted the native orientation of the mineral component, and the signal therefore tended toward isotropy. SAXS thickness measurements correlated well with more traditional diagenetic measurements obtained from FTIR spectroscopy. These parameters, infrared splitting factor (SF) and carbonate:phosphate ratio (C:P), were obtained for each sample following the method of Weiner and Bar-Yosef [14]. Both high SF and low C:P are thought to relate to increased crystallite perfection, larger size and reduced strain due to the loss of carbonate in the hydroxyapatite lattice. However, the correlation between SAXS and infrared measurements was only true for crystallites with thicknesses below 5 nm (thin-crystal samples). SAXS thickness measurements revealed a separate population of samples with enlarged crystallites, above 5 nm (thick-crystal samples). This thickening could not be simply explained by the presence of secondary mineral phases: while some of the thick-crystal samples contained calcite in the FTIR spectrum, just as many thin-crystal samples also showed traces of calcite.

Normally, bone crystallites are prevented from reaching thicknesses of 5 nm and above by the biological space limitation imposed by the collagen matrix; therefore, this increase in size reflects diagenetic alteration of the mineral, and the characteristics of the thick-crystal samples could be said to be diagenic, rather than biogenic. These thick-crystal samples did not show any related variation in either SF or C:P, that would indicate a reason for the increase in size; apparently, lattice perfection was not directly related to the substantial growth often observed in some samples. SF and C:P, as measurements of lattice perfection, can measure strain and composition in crystal lattices regardless of the crystal size, and hence may not reflect the full extent of diagenetic alteration to bone mineral. SAXS, on the other hand, provides a direct indication of the preservation or disruption of the biogenic crystal structure itself.

The presence of this population of crystals that have lost their biogenic structure, yet retain their lattice characteristics indicates that the crystal lattice perfection and retention of biogenic structure may only be weakly related in diagenesis (*cf.* Ref. [13]).

Two samples in particular had crystallites with thicknesses almost 10 times that of modern bone crystallites, which were beyond the limits of SAXS investigation within the current experimental system, and yet neither stood out from the rest of the samples in terms of SF or C:P. In our previous experience, crystals so enlarged have only been seen in experimentally cremated samples. For these samples, heating prior to burial or other similar postmortem treatment seems unlikely, since they both are Pleistocene cave bear (*Ursus spelaeus*) remains, albeit from different sites; the most likely cause is the presence of calcite in the samples, which was detected in the FTIR spectra. It is possible that these measurements of crystallites over 20 nm thick were caused by the limitation of the SAXS measurement itself. The range of crystallite thicknesses that theoretically can be measured by SAXS has been given as 0.5–50 nm [2]. However, at increasing thickness values, the angle required to measure crystallite parameters becomes smaller, and increasingly difficult to resolve from the non-scattered light absorbed by the primary beamstop. The crystal thicknesses of these samples may be measured more accurately using an increased sample to detector distance.

The remainder of the thick-crystal samples are more easily explained by the limitations of the sample set itself, rather than that of the apparatus or experimental procedure. All the archaeological bones used in this study have originated in cave sites, where apatite and carbonated apatite can often be precipitated as stable mineral elements [48]. In such an environment, parameters that measure lattice perfection may not change even though the crystallites themselves undergo profound alteration. In these samples, the difference between a lattice measurement, such as SF or C:P, and a real dimensional measurement such as

SAXS thickness, becomes clear. Alteration in apparently well-preserved, stable samples becomes evident when SAXS is used as a complementary technique.

### 2.1. SAXS and the prediction of organic preservation

The relationships between organic preservation – measured as weight percent nitrogen in whole bone powder (%N) – and the three mineral measurements (SAXS thickness, SF and C:P) were investigated. For all three mineral parameters, the samples were split into thin-crystal and thick-crystal populations. These populations could be considered biogenic (thin crystallites, close to modern thickness values) or diagenic (thick crystallites, most likely resulting from postmortem alteration). All three mineral measurements had a relatively weak relationship to remaining nitrogen in the sample; of these, SF had the closest correlation. Since, SAXS is a direct measurement of the alteration of biogenic structure, it will show changes that may not be reflected in lattice perfection indices. If the mineral surface and intact structure do indeed play a role in biomolecular preservation, then SAXS may be not only a better indicator of the true extent of mineral alteration, but also an "early-warning" system for the beginning of diagenetic change. For all three mineral measurements, a wide range of variation in mineral characteristics was seen at the end of the scale where no organic matrix remained. This indicated that, once the regulation of the collagenous component was lost, the mineral was free to change: lattice, shape and thickness – all shifted to the most stable configuration available in the burial microenvironment.

A small subset of samples that were known positives for ancient DNA ($n$ = 12) was included in the overall sample set. These were all thin-crystal samples, which reinforces the theory that DNA may be bound to the mineral surface; if crystals thicken substantially and the mineral surface is altered, then the DNA may be lost. The DNA positives showed much stronger relationships between mineral alteration and organic preservation than the remainder of the samples, the strongest of which was a direct relationship between crystal thickness and %N. This points to a specific diagenetic "path" that samples must remain on to have enough DNA surviving for amplification; the best samples have little or no mineral alteration and nearly modern levels of nitrogen remaining, but a certain amount of mineral change and nitrogen loss is allowed as long as the two remain closely related. Eventually, a stage is reached at which too much DNA has been lost or degraded to be amplifiable.

This preliminary work has shown evidence of a link between alteration in crystallite thickness and shape, and the preservation of organic material in bone. In future, a similar study using bones from younger open sites, where mineral diagenesis follows a different pathway, will establish more clearly the use of SAXS measurements to determine preservation and retrievability of organic material. The link found here between crystal thickness and other measures of mineral alteration has also reinforced conclusions reached in earlier work involving a microfocus SAXS technique [11].

Previously, it had been postulated that, the alteration to mineral surfaces observed using SAXS was linked to loss of organic material and changes to SF measurements [11]. While this previous work was able to determine crystal shape and size using SAXS over small areas within a bone cross section, there was no similar microanalysis technique to examine the corresponding infrared spectrum or organic residue present at these interior sites.

The development of a bulk technique for examining crystal shape and thickness in conjunction with well-established infrared spectroscopy and elemental analysis techniques has lent added confidence to other works on intra-bone variability of preservation.

## 3. MICROFOCUS SAXS AND TWO-DIMENSIONAL MAPPING

The development of microfocus lenses at synchrotron light sources has facilitated the analysis of biological composite materials on the nanoscopic level. Beamline ID18F at the European Synchrotron Radiation Facility (ESRF, Grenoble, France) has developed an X-ray microprobe based on compound refractive lenses [49], capable of focusing an X-ray beam down to $1.5 \times 15$ μm [50,51] which can be used for microfocus small-angle X-ray scattering (μSAXS) analysis of bone thin sections.

Heterogeneity in archaeological bone mineral tends to increase postmortem, which in turn may contribute to the loss of biomolecular information. Previous μSAXS work on archaeological bone [11] has shown areas of intact, unaltered bone crystallites in regions immediately adjacent to mineral that has undergone severe diagenetic change. Currently, μSAXS has been used to produce two-dimensional maps of bone crystallite properties, which can not only detect the extent of localised variation in modern bone but also record heterogeneity in archaeological samples.

Sections approximately 200 μm thick were cut from unembedded bone using a diamond annular microtome (SP1600, Leica Microsystems, Germany) under constant aqueous irrigation to prevent heating. μSAXS experiments were conducted on beamline ID18F at the ESRF. Removal of the microfocus apparatus from the X-ray beam path caused a sub-millimetre region of the bone to be bathed in the direct beam, and the transmission image resulting from differential absorbance in the sample enabled osteological features to be identified in the region selected. This transmission camera image was used to find an area of interest, either with intact histological features or signs of microbial damage. The microfocus lens was then replaced, and a 200-μm square raster scan of the selected area was carried out, with a scattering measurement taken every 10 μm. Data was processed using software developed in-house, as described previously [11], and values for crystal thickness in nanometers ($T$) and a shape parameter ($\eta$) were obtained from each scattering measurement. A "mesh" image, composed of separate but related scattering measurements across a delineated topological area, was then produced by ranking the values for thickness and shape obtained for each 10-μm step, and generating a greyscale image progressing from lighter (thinner or more polydisperse crystals) to darker (thicker or more needle-like crystals) areas. A mean value, standard deviation, and coefficient of variance were also determined for $T$ and $\eta$ over each 200-μm area.

The two-dimensional mesh scan areas produced similar results as were evident in previous one-dimensional scanning experiments [11]. The modern control sample had mean crystallite thickness values within the range for fresh bone (defined as 2.8–3.8 nm [2]); the shape parameter suggested plate-like crystallites. In the archaeological samples, values for $T$ and $\eta$ were more diverse than the modern sample, both in terms of real values and in the amount of variation present in the scan. It is apparent that diagenetic alteration to the bone mineral can result in thinner or thicker crystallites than those seen physiologically, but the disparity between these overall values as well as altered variation across the scan is indicative of the state of preservation of the sample. Diagenetic processes appear to cause both thickening and thinning of crystallites, as well as shape change. This indicates that alteration can stimulate more than one remodelling regime in the archaeological bone mineral. The archaeological samples used in this study were of widely differing ages and provenances, which implies that the nanotextural variation observed was not due to a particular site environment or duration of deposition.

The effect of histological features on the variation observed is clear in the two-dimensional mesh images. In the mesh images of a modern human bone (MHS1) shown in Fig. 7, the concentric lamellae surrounding the secondary osteon at the right side of the transmission radiograph is clearly visible, as it is the contrast between the bone crystallites in the secondary osteon and the crystallites present in the primary osteon on the left that appear to have been remodelled as part of normal bone turnover. Variation is greater in shape than in thickness, as would be expected if the shape measurement can be affected by crystallite orientation. The orientation of the scattering pattern may contribute to the shape parameter due to the small volume of bone sampled in each measurement; the extent of this effect

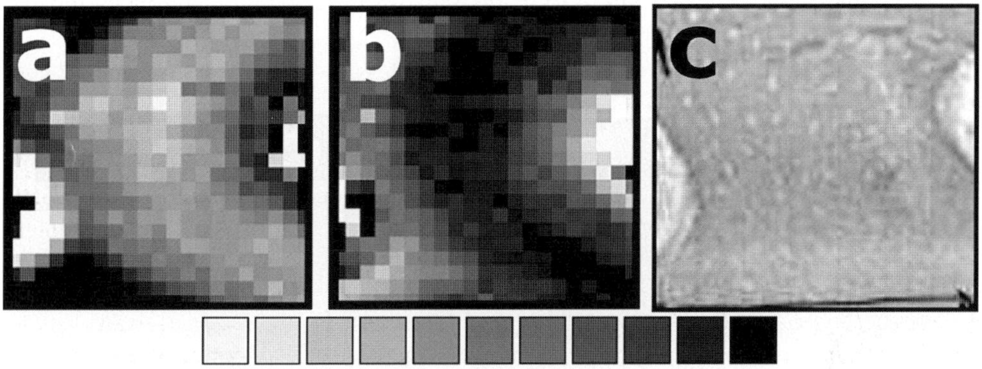


**Fig. 7.** Raster images and transmission radiograph of modern human bone. Values for crystal thickness of the smallest dimension in nanometers ($T$) and a shape parameter ($\eta$) were obtained from each scattering measurement. Panel **a**, crystallite shape measurements (lighter areas contain more polydisperse crystallites, darker areas contain more needle-like crystals); **b**, crystallite thickness (lighter areas correspond to thinner crystallites); **c**, radiograph showing histological features. Two osteons are visible on either side of the transmission radiograph, which appear as bright spots in the thickness and shape mesh images. The concentric rings of bone around the osteonal centre on the right are also clearly visible in the centre of the mesh images, as a curving band of thicker, platelike crystallites. The left osteon is surrounded by thinner, more needle-like crystallites, which the bone surrounding the right osteon seems to cut through; this appears to represent a primary (left) and secondary (right) osteon structure. The osteonal centres themselves show up as bright centres of polydisperse, needle-like crystallites; this probably reflects the distortion in the scattering results arising from the void of the Haversian canal. A greyscale bar is provided at the bottom of the image, illustrating the progression from light (polydisperse $\eta$, thin $T$) to dark (needle-like $\eta$, thick $T$) areas.

is unclear, however, and will form the basis for a future study. Crystallite thickness measurements should not be affected by the shape or orientation, and consequently this measurement is more consistent across the scan area.

Sample BP-3 (Fig. 8) is an archaeological aurochs (primordial European cow, *Bos primigenius*) sample from the United Kingdom, dated to 5936 ± 34 radiocarbon years before present (bp). The sample had a high level of residual nitrogen and little nanotextural alteration compared to the modern sample examined here. The crystallite shape is consistently needle-like, with variation across the scan similar to that seen in modern samples; the thickness values are close to those in modern cow bone. The retrieval of ancient DNA in at least two independent attempts [52] speaks to the excellent preservational condition of this sample, including at the nanotextural level.

Sequences of both mitochondrial and nuclear (cytochrome *c*) DNA were obtained from sample SP-015 (Fig. 9). SP-015 is an archaeological cave bear (*Ursus spelaeus*) sample from France, dated to 25 000–35 000 years bp [53]. This sample also had primarily needle-like crystals close to modern thickness, reflecting good preservation on the nanotextural level. The surface of the apatite crystallites is not considerably altered, allowing for the preservation of amplifiable ancient DNA in microniches, or areas of mineral sheltering that

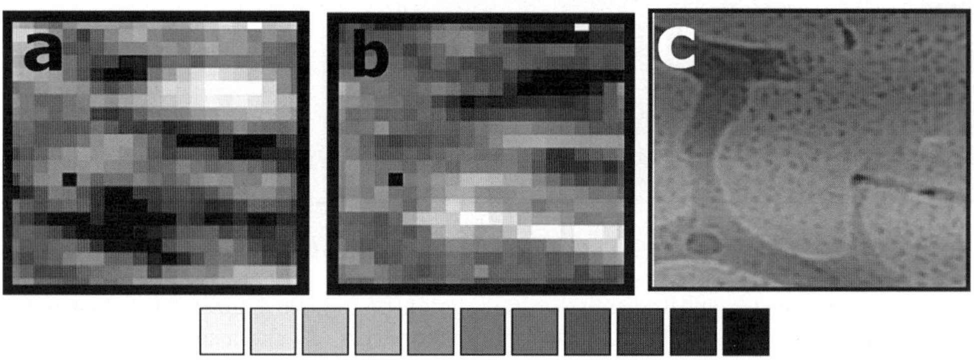


**Fig. 8.** Raster images and transmission radiograph of well-preserved archaeological aurochs (primordial European cow) bone (BP-3); maps are the same as defined in Fig. 7. Panel **a**, crystallite shape; **b**, crystallite thickness; **c**, radiograph showing histological features. From the transmission camera image of BP-3, it is apparent that the bone is in longitudinal section rather than transverse section. A channel runs through the bone in the lower right quadrant of the radiograph. This is reflected in the area of thin, plate-like crystals observed in the corresponding mesh scans, which again may mislead scattering results similar to the ones observed with the osteonal centres in the modern mesh scans. Overall, the BP-3 mesh scans are somewhat striated in appearance, reflecting the longitudinal histology of the section, and are fairly uniform despite the presence of histological vacuities similar to the osteonal centres seen in the modern samples. A greyscale bar is provided at the bottom of the image, illustrating the progression from light (polydisperse $\eta$, thin $T$) to dark (needle-like $\eta$, thick $T$) areas.

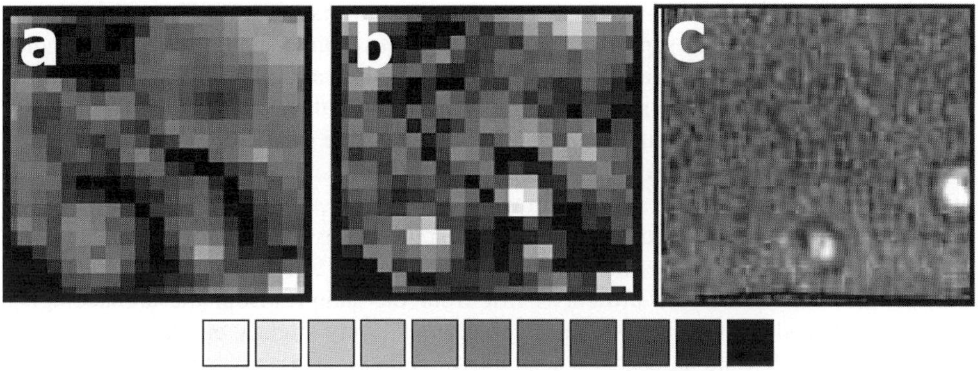


**Fig. 9.** Raster images and transmission radiograph of poorly-preserved archaeological cave bear bone (SP-015); maps are the same as defined in Fig. 7. Panel **a**, crystallite shape; **b**, crystallite thickness; **c**, radiograph showing histological features. The transmission radiograph of SP-015 is grainy, but clearly shows two osteonal centres in the lower right corner of the image. While offset slightly from the transmission image, the mesh scans reproduce the histological structure of the bone, including the two osteonal centres (visible as bright spots of relatively thin crystals in the lower right-hand part of the thickness map). A greyscale bar is provided at the bottom of the image, illustrating the progression from light (polydisperse $\eta$, thin $T$) to dark (needle-like $\eta$, thick $T$) areas.

allow for long-term DNA survival despite the loss of integrity in other diagenetic indicators, as postulated by some ancient DNA researchers [54].

The results from μSAXS experiments show that nanotextural variation is present across areas of modern, unaltered bone. While these effects are real, partial sampling or changes in the volume fraction may affect the calculated thickness or shape values. The extent of this is not yet fully understood. Additionally, the changes in orientation that occur around osteons in histologically intact thin sections may cause variation in the shape measurements, although this is considered to be negligible due to the spherical averaging step in data analysis. Knowledge of the level of nanotextural variability in bone unaltered by diagenetic processes is crucial to understanding the contribution of diagenesis to crystallite alteration in archaeological samples.

Ancient DNA amplification was reported from both the archaeological samples. Differences in age and sample origin, however, could have affected preservation. BP-3 was reported to be successful in two or more independent ancient DNA extractions [52]. SP-015, while considerably older than BP-3, yielded very high amounts of both mitochondrial and nuclear ancient DNA, enough for a complete cytochrome *b* sequence. Limited alteration to the nanostructure, coupled with reduced variation in crystallite characteristics, could explain why, despite its age, this sample yielded amplifiable DNA. Apatite has a tendency to bind free DNA [47], and this may facilitate its survival.

A link between intact mineral structure and the preservation of ancient DNA does argue for the possibility of "microniches" in the bone, as described by an earlier study [54]. In previous work [11], it was shown in one-dimensional scans that microbial alteration caused areas of nanotextural disruption that could coexist next to areas of intact, seemingly unaltered bone mineral with barely 10 μm separating the two. In the two-dimensional scan areas studied here, microbial alteration was not present; rather, the path of alteration seems to be an early shape change to increasingly needle-like crystallites, with the initial loss of some of the biogenic protein. Variation in the shape parameter across the archaeological scans remains similar or is reduced compared to that seen in the modern scan, implying that the crystallites change to a more stable shape following protein loss, but that the variation due to orientation or histological voids is conserved, despite the loss of histological structure.

Following this initial shape change, crystallites have the freedom to enlarge across the smallest dimension, as more protein is lost from the sample. Both the archaeological samples contained reduced nanotextural variation and only slightly enlarged crystallites, and were both high-yield ancient DNA samples. While SP-015 had nearly lost double the nitrogen than the BP-3, the crystallites had stabilised at a slightly thickened needle-like shape; we can speculate that this may have sheltered the DNA from further loss. BP-3 showed little alteration in any parameter examined here, including nanotexture, and the sample is therefore as promising as possible for preservation of ancient DNA. If the binding of DNA fragments to apatite surfaces occurs without subsequent mineral alteration, then the DNA could be quite well protected in a sort of mineral carapace. This protection could explain the preservation of ancient DNA in the more highly altered SP-015. A link between ancient DNA preservation and overall crystallite characteristics observed in powdered samples has recently been elucidated [28], which is reinforced by the links observed here over intact bone thin sections.

It is evident that two-dimensional mapping of modern and archaeological bone specimens can reveal localised variation due to histological features as well as diagenetic effects. The existence of microniches of preservation, in which biogenic mineral, and hence viable biomolecular material remains intact, may be visualised using this technique in the future. This method represents a cutting edge in the X-ray microfocus analysis of archaeological bone. A more thorough test of the molecular niche hypothesis would ideally be done in conjunction with a microanalysis method of testing for the presence of DNA, such as histological staining [55], molecular hybridisation [56], or *in-situ* PCR; nevertheless, many technical constraints would need to be overcome for this to be possible.

## 4. DETECTION OF BURNING AND CREMATION

The ability to identify burning and burned bone in the forensic and archaeological records has long been an important and contentious issue. Several techniques to determine burning or heating regimen used in archaeological contexts have been derived, with varying levels of success [57–60]. Determination of the temperature and duration of burning, as well as the background noise of potential diagenetic effects [61] would shed light on cooking practices, the early use of fire, cremation as a burial rite, and other archaeological and paleoanthropological puzzles. Further, the effects of burning on bone specimens and the determination of the techniques used are crucial in the resolution of forensic cases where cremation or other fire damage to remains is present [62–66]. Changes to the biogenic composition and structure of the bone mineral following heat treatment at different

**Table 1.** Crystal thickness values from SAXS profiles

| Sample code | Temperature (°C) | Time (min) | $T$ (nm) |
|---|---|---|---|
| 3-0002-05 | 500 | 15 | 5.24 |
| 3-0002-07 | 500 | 15 | 5.65 |
| 3-0003-02 | 700 | 15 | 10.37 |
| 3-0003-06 | 700 | 15 | 14.09 |
| 3-0004-03 | 900 | 15 | 17.49 |
| 3-0004-05 | 900 | 15 | 22.59 |
| 3-0005-01 | 500 | 45 | 7.81 |
| 3-0005-05 | 500 | 45 | 6.71 |
| 3-0006-01 | 700 | 45 | 16.11 |
| 3-0006-02 | 700 | 45 | 15.60 |
| 3-0007-02 | 900 | 45 | 31.26 |
| 3-0007-10 | 900 | 45 | 26.66 |
| Control 1 | N/A | N/A | 2.79 |
| Control 2 | N/A | N/A | 2.36 |

temperatures could be used to simulate burning scenarios. It would be valuable to know the temperature at which crystallites begin to change, how rapid the alteration can be, and whether any additional factors such as age or sex of the animal can affect this process.

Previously, X-ray diffraction (XRD) and Fourier-transform infrared spectroscopy (FTIR) have been used to determine changes to the mineral phase of bone during heating or burning [57,60,64,67–69]. A generalised trend has been observed toward a more "perfect" or "crystalline" phase of hydroxyapatite at temperatures up to 1000°C; above this, the emergence of different mineral phases can sometimes be discerned. However, fine-scale changes to bone ultrastructure at temperatures below 1000°C can be difficult to detect using XRD [70,71]. Recent results have elucidated more clearly the changes to bone mineral during burning using a combination of XRD (or wide angle-X-ray scattering, WAXS) and SAXS techniques. Changes to crystallite size and shape during early stages of burning and at lower temperatures are more readily visible using SAXS, thus opening up a new route into the investigation of the effects of heat treatment on bone mineral in biomaterials research as well as archaeological and forensic contexts.

Samples of fresh cortical bone removed from sheep long bones were defleshed and heated experimentally to 500, 700 and 900°C for either 15 or 45 min. Pairs of samples were subjected to the same heating regimen, and 14 samples, including two controls, were analysed. Whole long bone samples were heated in pairs in an electric muffle furnace. The samples were placed on heatproof ceramic trays in order to aid retrieval after heating. Once the temperature had reached 200°C, each pair of long bones was placed into the furnace and allowed to heat up to the designated temperature. This was deemed more akin to natural burning or heating situations where soft tissue insulates the bone surface in the early stages of heating. Preheating at 200°C removes the potential impact of extremely rapid heating as an influence on hard tissue microstructure [72]. Once the chosen temperature was reached, the samples were retained at that temperature for 15 or 45 min. After this, they were removed from the furnace and allowed to cool naturally before being handled again. Table 1 lists the crystal thickness results from these heating experiments.

The two control samples have crystallite thicknesses of approximately 2.3–2.8 nm, which fall slightly below the average values for crystallites in mature faunal bone [25,27]. With heating, the crystallites grow substantially in size, from just over 5 nm in the samples heated at 500°C for only 15 min, to over 30 nm in samples heated at 900°C for 45 min. The thickness values increased for longer heating times, but a substantial change has been wrought during the first 15 min at high temperature. With prolonged heating, the difference between the samples within a pair is reduced compared to the differences between pairs, indicating that there may be a stable crystal size for a specific temperature. Samples heated up to 900°C displayed such high levels of alteration, which resulted in crystal thickness increases of up to tenfold, that these were rerun for 9-h exposures; these results are shown in Table 2. The 3-h runs produced very weak scattering data for these four samples. It appears that with increasing crystal size, the scattering profiles become weaker and less informative, as the necessary incident angle for sufficient X-ray scattering becomes smaller when the incident beam is scattered at such a low angle that the resultant X-rays are increasingly absorbed into the beamstop. From the longer runs, similar, although slightly larger, thickness values were obtained for the samples heated for 45 min. The samples heated for 15 min produced anomalous results due to the limitations of the technique, as large crystallites become increasingly difficult to measure accurately.

Crystal shape profiles were also determined for the 14 samples, and are shown in Figs. 10–12. The plots were corrected for thickness variation following a procedure detailed in Ref. [25], allowing a more direct comparison of crystallite habit alone. Figure 10 displays the even needle morphology seen in the unheated control samples. In Fig. 11, changes in shape after 15 min of heating are evident. The pair of samples heated to 500°C shows a more plate-like shape, whereas the considerably larger crystals in the samples heated to 700°C have a polydisperse morphology. This may reflect the difficulty in calculating the shape of crystals so large using this method. After 45 min of heating (Fig. 12), the shapes of the samples heated to 500°C are still plate-like, although the two curves are much more similar to each other. The samples heated to 700°C show a similar trend: still polydisperse, but more alike. The samples heated to 900°C are not shown; due to their large size and weak scattering trait, the crystal habit could not be determined.

From these results, it appears that significant changes in the crystallite shape and thickness occur during experimental heating. Wide-angle X-ray scattering (WAXS) measurements confirmed earlier results that found increasingly crystalline hydroxyapatite at the temperatures

**Table 2.** Crystal thicknesses measured over 9 h for samples heated at 900°C

| Sample code | Temperature (°C) | Time (min) | *T* (nm) |
|---|---|---|---|
| 3-0004-03 | 900 | 15 | 29.39 |
| 3-0004-05 | 900 | 15 | 74.04 |
| 3-0007-02 | 900 | 45 | 31.83 |
| 3-0007-10 | 900 | 45 | 32.90 |

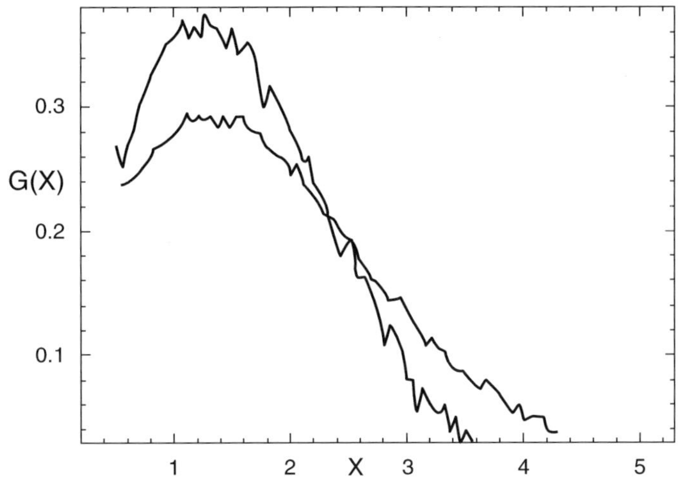


**Fig. 10.** Thickness-corrected plots illustrating needle-like morphology in both control (unheated) samples.

used here, but detected no new mineral phases. Previous studies of heated bone have found calcium oxide (CaO) formation at temperatures above 700°C [69], but the primary effect of heating is to generate larger and more crystalline hydroxyapatite [57]. It has been suggested that the formation of different mineral phases as a result of heating may be a function of age: CaO has been found in human samples older than 22 years in one study [71] and a link between skeletal maturity and mineral change during heating has been

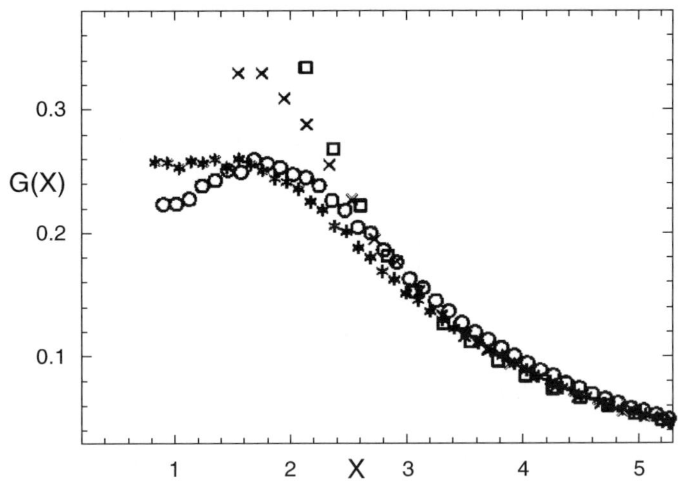


**Fig. 11.** Thickness-corrected plots for pairs of samples heated for 15 min at 500°C (stars and circles) or 700°C (squares and crosses). The development of a more plate-like habit at the lower temperature, progressing to polydisperse crystals as the heat increases, is evident. (After Ref. [29].)

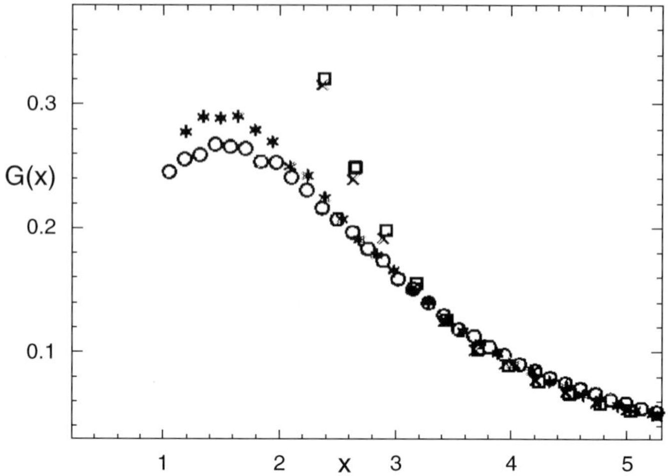


**Fig. 12.** Thickness-corrected plots for pairs of samples heated for 45 min at 500°C (stars and circles) or 700°C (squares and crosses). The curves again show plate-like crystals in the lower temperature and polydisperse crystals in the higher temperature samples. (After Ref. [29].)

found to be common to several mammalian species [68]. Here, results show a slight narrowing of peaks with increasing heat, but contain no clear evidence of new mineral formation, corroborating the earlier conclusions that the existing hydroxyapatite becomes more crystalline with heat.

Small-angle X-ray scattering (SAXS) results show an increase in thickness and an alteration in crystal morphology with heat, which correlates with earlier electron microscopic investigations into heated bone structure. Raspanti *et al.* [73], and more recently Quatrehomme *et al.* [74], using scanning electron microscopy showed that, there was little structural change to bone heated to 500°C, but on heating up to 700°C or higher, the mineral phase was replaced by large clumps of crystallites. A similar change is reflected in the mineral alteration seen in the SAXS profiles, albeit on a different scale; larger crystallites of indeterminate polydisperse habit appeared upon heating to 700°C, and heating up to 900°C made the measurement of crystallite size or shape using SAXS difficult at best. To determine more accurately the characteristics of the crystallites in samples heated past 700°C, a longer sample-to-detector distance, such as those found at synchrotron light sources, could perhaps be used, which may delineate the structure of these large crystallites more clearly.

Shape changes in crystallites and an initial thickening were evident in the samples examined here in the first 15 min of heating. With increased heating time, the shape alteration remained very similar to that obtained in 15 min, but the differences between the samples in each pair were slightly reduced. Crystals became thicker after 45 min of heating, and the differences between pairs became more pronounced than the differences within pairs. This implies that there is a temperature-specific stable mineral structure that emerges gradually with increased heating time: the first 15 min allow for initial shape change, while after 45 min the thickness increases without much additional shape change. This may be due to a sintering process that produces hydroxyapatite crystals of a particular shape and size following specific heating regimens.

The crystal change apparent in these SAXS measurements is not as easily discernible as in XRD traces. Corresponding WAXS measurements (data not shown) contained little to differentiate them aside from slight peak narrowing and splitting, while the crystallites were growing substantially and changing habit. Unlike XRD, SAXS provides direct measurements of crystallite size and shape that are independent of the perfection of the crystal lattice [25]. However, the pattern of crystal change from small imperfect crystallites to larger, more perfect ones is reinforced by both sets of measurements, verifying the usefulness of SAXS as a complementary technique to study bone mineral structure and change resulting from heat treatment. With a greater range of samples and longer heating times, it may be possible to discern the precise characteristics of crystals heated to a specific temperature or for a specific time. This additional information would make SAXS valuable in establishing effective screening techniques for the generation of biogenic apatite with minimal alteration to mineral structure for osteoimplantation. It could also act as a simple method, requiring minimal sample preparation, for the tracing of heating regimens or exposures encountered in forensic or archaeological contexts.

## 5. CONCLUSIONS

Alteration in the mineral phase was observed to be significant for the preservation of biomolecules in bone samples, particularly in terms of ancient DNA. Small-angle X-ray scattering focuses on the structural dimensions of the crystallites present in bone, thus measuring the bone nanotexture. A link was demonstrated between alteration to crystal structure (in terms of thickness or shape) and other diagenetic changes, including loss of nitrogenous material as well as the currently used measures of mineral alteration. It was found that SAXS could reveal alteration to bone crystallite surfaces that could occur without corresponding changes in the lattice composition or strain. The measure of crystallite thickness was also seen to relate to the preservation of protein as measured by percent nitrogen, with about the same strength of correlation as that seen between nitrogen level and splitting factor. For DNA positive samples, a stronger correlation between crystallite thickness and organic preservation was observed than that seen for carbonate content or splitting factor. Thus, SAXS revealed a common feature of the DNA positive samples that could provide an explanation for the preservation of amplifiable sequences. Retention of the biogenic crystal surface may allow small polar molecules such as polynucleotides and non-collagenous proteins such as osteocalcin to remain bound to the apatite, thus partially sheltering from degradation.

Microfocus methods were used as a means to explore further the retention of pockets of bone with biogenically intact mineral in archaeological samples, for testing the theory of microniches of preservation first put forth by Geigl [54]. Since earlier results had shown a potential link between crystallites with biogenic characteristics and the retention of biomolecular material, it seemed possible that biogenic crystallites surviving even in altered bone samples could act as havens for small polynucleotides. This was reinforced by earlier results using μSAXS [11], which showed that the regions of microbial degradation harboured altered crystallites, but these could be surrounded by histologically unaltered areas with biogenic crystallites. The results presented here are the first examples of a two-dimensional map of an area of bone based on crystallite characteristics. Further innovation, hopefully, will lead to the production of high-resolution maps of archaeological bone, in which microniches of bone crystallites can be visualised directly.

Furthermore, SAXS provides evidence complementary to that generated by traditional XRD or WAXS in the characterisation of heated bone. However, fine-scale changes in crystallite size and shape that are not measured directly using XRD are readily elucidated using SAXS, however and therefore changes in the crystal structure that may not be readily apparent otherwise become more clear. We are confident that the techniques described here can be honed for use as a more accurate determinant of crystallite change during heating, thus providing an additional means of determining the effects of heat treatment on biogenic hydroxyapatite or tracing burning practices in the forensic and archaeological records.

## REFERENCES


[1] R.Z. LeGeros, Apatites in biological systems, *Progress in Crystal Growth and Characterisation*, **4** (1981), 1–45.

[2] P. Fratzl, S. Schreiber, K. Klaushofer, Bone mineralization as studied by small-angle X-ray scattering. *Connective Tissue Research*, **34** (1996a) 4, 247–254.

[3] P. Fratzl, Statistical model of the habit and arrangement of mineral crystals in the collagen of bone, *Journal of Statistical Physics*, **77** (1994) 1–2, 125–143.

[4] P. Fratzl, N. Fratzl-Zelman, K. Klaushofer, G. Vogl, K. Koller, Nucleation and growth of mineral crystals in bone studied by small-angle X-ray scattering, *Calcified Tissue International*, **48** (1991), 407–413.

[5] P. Fratzl, M. Groschner, G. Vogl, H. Plenk, J. Eschberger, N. Fratzl-Zelman, K. Koller, K. Klaushofer, Mineral crystals in calcified tissues: a comparative study by SAXS, *Journal of Bone and Mineral Research*, **7** (1992) 3, 329–334.

[6] H.-R. Wenk, F. Heidelbach, Crystal alignment of carbonated apatite in bone and calcified tendon: results from quantitative texture analysis, *Bone*, **24** (1999) 4, 361–369.

[7] S. Weiner, W. Traub, Bone structure: from angstroms to microns, *FASEB Journal*, **6** (1992), 879–885.

[8] M.M. Giraud-Guille, Plywood structures in nature, *Current Opinion in Solid State & Materials Science*, **3** (1998), 221–227.

[9] M.J. Collins, C.M. Nielsen-Marsh, J. Hiller, C.I. Smith, J.P. Roberts, R.V. Prigodich, T.J. Wess, J. Csapò, A.R. Millard, G. Turner-Walker, The survival of organic matter in bone: a review, *Archaeometry*, **44** (2002) 3, 383–394.

[10] C.J. Hackett, Microscopic focal destruction (tunnels) in exhumed human bones, *Medicine, Science, and the Law*, **21** (1981), 243–265.

[11] T. Wess, I. Alberts, J. Hiller, M. Drakopoulos, A.T. Chamberlain, M. Collins, Microfocus small-angle X-ray scattering reveals structural features in archaeological bone samples: detection of changes in bone mineral habit and size, *Calcified Tissue International*, **70** (2002) 2, 103–110.

[12] L.M. Miller, R. Huang, M.R. Chance, C.S. Carlson, Applications of fluorescence-assisted infrared micro-spectroscopy to the study of osteoporosis, *Synchrotron Radiation News*, **12** (1999) 1, 21–28.

[13] L.M. Miller, V. Vairavamurthy, M.R. Chance, R. Mendelsohn, E.P. Paschalis, F. Betts, A.L. Boskey, In situ analysis of mineral content and crystallinity in bone using infrared micro-spectroscopy of the $\nu_4$ $PO_4^{3-}$ vibration, *Biochimica et Biophysica Acta*, **1527** (2001), 11–19.

[14] S. Weiner, O. Bar-Yosef, States of preservation of bones from prehistoric sites in the Near East: a survey, *Journal of Archaeological Science*, **17** (1990), 187–196.

[15] L.E. Wright, H.P. Schwarcz, Infrared and isotopic evidence for diagenesis of bone apatite at Dos Pilas, Guatemala: palaeodietary implications, *Journal of Archaeological Science*, **23** (1996) 6, 933–944.

[16] M.J. Farquharson, R.D. Speller, M. Brickley, Measuring bone mineral density in archaeological bone using energy-dispersive low angle X-ray scattering techniques, *Journal of Archaeological Science*, **24** (1997), 765–772.

[17] A. Person, H. Bocherens, J.-F. Saliege, F. Paris, V. Zeitoun, M. Gerard, Early diagenetic evolution of bone phosphate: an X-ray diffractometry analysis, *Journal of Archaeological Science*, **22** (1995), 211–221.

[18] I. Reiche, L. Favre-Quattropani, T. Calligaro, J. Salomon, H. Bocherens, L. Charlet, M. Menu, Trace element composition of archaeological bones and post-mortem alteration in the burial environment, *Nuclear Instruments and Methods in Physics Research B*, **150** (1999), 656–662.

[19] A. Bartsiokas, A.P. Middleton, Characterization and dating of recent and fossil bone by X-ray diffraction, *Journal of Archaeological Science*, **19** (1992), 63–72.

[20] J.D. Termine, A.S. Posner, Infrared determination of the percentage of crystallinity in apatitic calcium phosphates, *Nature*, **211** (1966), 268–270.

[21] A. Sillen, A. Parkington, Diagenesis of bone from Eland's Bay Cave, *Journal of Archaeological Science*, **23** (1996), 535–542.

[22] A. Götherström, M.J. Collins, A. Angerbjörn, K. Lidén, Bone preservation and DNA amplification, *Archaeometry*, **44** (2002) 3, 395–404.

[23] T.A. Surovell, M.C. Stiner, Standardizing infrared measures of bone mineral crystallinity: an experimental approach, *Journal of Archaeological Science*, **28** (2001), 633–642.

[24] E. Wachtel, S. Weiner, Small-angle X-ray scattering study of dispersed crystals from bone and tendon, *Journal of Bone and Mineral Research*, **9** (1994) 10, 1651–1655.

[25] P. Fratzl, S. Schreiber, A. Boyde, Characterization of bone mineral crystals in horse radius by small-angle X-ray scattering, *Calcified Tissue International*, **58** (1996b), 341–346.

[26] N.P. Camacho, S. Rinnerthaler, E.P. Paschalis, R. Mendelsohn, A.L. Boskey, P. Fratzl, Complementary information on bone ultrastructure from scanning small-angle X-ray scattering and Fourier-transform infrared microspectroscopy, *Bone*, **25** (1999) 3, 287–293.

[27] T.J. Wess, M. Drakopoulos, A. Snigirev, J. Wouters, O. Paris, P. Fratzl, M. Collins, J. Hiller, K. Nielsen, The use of small-angle X-ray diffraction studies for the analysis of structural features in archaeological samples, *Archaeometry*, **43** (2001) 1, 117–129.

[28] J.C. Hiller, M.J. Collins, A.T. Chamberlain, T.J. Wess, Small-angle X-ray scattering: a high-throughput method for investigating archaeological bone preservation, *Journal of Archaeological Science*, **31** (2004), 1349–1359.

[29] J.C. Hiller, T.J.U. Thompson, M.P. Evison, A.T. Chamberlain, T.J. Wess, Bone mineral change during experimental heating: an X-ray scattering investigation, *Biomaterials*, **24** (2003), 5091–5097.

[30] M. Parker-Pearson, A. Chamberlain, O. Craig, P. Marshall, J. Mulville, H. Smith, C. Chenery, M. Collins, G. Cook, G. Craig, J. Evans, J. Hiller, J. Montgomery, J.-L. Schwenninger, G. Taylor, T. Wess, Evidence for mummification in prehistoric Britain, *Antiquity*, **79** (2005), 529–546.

[31] A. Guinier, G. Fournet, *Small-angle scattering of X-rays*, Wiley, New York, 1955.

[32] O. Glatter, O. Kratky, *Small-angle X-ray scattering*, Academic Press, London, 1982.

[33] N. Matsushima, M. Akiyama, Y. Terayama, Quantitative analysis of small-angle X-ray scattering of bone: determination of sizes of its collagen and apatite components, *Japanese Journal of Applied Physics*, **20** (1981) 4, 699–702.

[34] D. Halliday, R. Resnick, J. Walker, *Fundamentals of Physics*, fourth Ed., John Wiley and Sons, New York, 1993.

[35] O. Kratky, A survey, in: O. Glatter, O. Kratky (Eds.), *Small-angle X-ray scattering*, Academic Press, London, 1982.

[36] J.J. Thomas, H.M. Jennings, A.J. Allen, The surface area of cement paste as measured by neutron scattering: evidence for two C-S-H morphologies, *Cement and Concrete Research*, **28** (1998) 6, 897–905.

[37] G. Porod, General theory, in: O. Glatter, O. Kratky, (Eds.), *Small-angle X-ray scattering*, Academic Press, London, 1982.

[38] N. Matsushima, M. Akiyama, Y. Terayama, Y. Izumi, Y. Miyake, The morphology of bone mineral as revealed by small-angle X-ray scattering, *Biochimica et Biophysica Acta*, **801** (1984), 298–305.

[39] P. Fratzl, H.S. Gupta, O. Paris, A. Valenta, P. Roschger, K. Klaushofer, Diffracting 'stacks of cards': some thoughts about small-angle scattering from bone, *Progress in Colloid and Polymer Science*, **130** (2005), 1–7.

[40] I.B. Colson, J.F. Bailey, M. Vercauteren, B.C. Sykes, R.E.M. Hedges, The preservation of ancient DNA and bone diagenesis, *Ancient Biomolecules*, **1** (1997) 2, 109–117.

[41] S. Haynes, J.B. Searle, A. Bretman, K.M. Dobney, Bone preservation and ancient DNA: the application of screening methods for predicting DNA survival, *Journal of Archaeological Science*, **29** (2002), 585–592.

[42] J.L. Bada, X.S. Wang, H. Hamilton, Preservation of key biomolecules in the fossil record: current knowledge and future challenges. *Philosophical Transactions of the Royal Society of London Series B*, **354** (1999), 77–87.

[43] H.N. Poinar, B.A. Stankiewicz, Protein preservation and DNA retrieval from ancient tissues, in: *Proceedings of the National Academy of Sciences of the USA*, **96** (1999), 8426–8431.

[44] A. Cooper, Reply to Stoneking: Ancient DNA – how do you really know when you have it? *American Journal of Human Genetics*, **60** (1997), 1001–1002.

[45] A. Cooper, H.N. Poinar, Ancient DNA: do it right or not at all, *Science*, **289** (2000), 1139.

[46] A.S. Posner, The structure of bone apatite surfaces, *Journal of Biomedical Materials Research*, **19** (1985), 241–250.

[47] M. Okazaki, Y. Yoshida, S. Yamaguchi, M. Kaneno, J.C. Elliott, Affinity binding phenomena of DNA onto apatite crystals, *Biomaterials*, **22** (2001), 2459–2464.

[48] P. Karkanas, O. Bar-Yosef, P. Goldberg, S. Weiner, Diagenesis in prehistoric caves: the use of minerals that form *in situ* to assess the completeness of the archaeological record, *Journal of Archaeological Science*, **27** (2000), 915–929.

[49] B. Lengeler, C.G. Schroer, M. Richwin, J. Tummler, M. Drakopoulos, A. Snigirev, I. Snigireva, A microscope for hard X-rays based on parabolic compound refractive lenses, *Applied Physics Letters*, **74** (1999b), 3924–3926.

[50] A. Snigirev, V. Kohn, I. Snigireva, B. Lengeler, A compound refractive lens for focusing high-energy X-rays, *Nature*, **384** (1996), 49–51.

[51] B. Lengeler, C. Schroer, J. Tummler, B. Benner, M. Richwin, A. Snigirev, I. Snigireva, M. Drakopoulos, Imaging by parabolic refractive lenses in the hard X-ray range, *Journal of Synchrotron Radiation*, **6** (1999a), 1153–1167.

[52] C.S. Troy, D.E. MacHugh, J.F. Bailey, D.A. Magee, R.T. Loftus, P. Cunningham, A.T. Chamberlain, B.C. Sykes, D.G. Bradley, Genetic evidence for Near Eastern origins of European cattle, *Nature*, **410** (2001), 1088–1091.

[53] O. Loreille, L. Orlando, M. Patou-Mathis, M. Philippe, P. Taberlet, C. Hanni, Ancient DNA analysis reveals divergence of the cave bear, *Ursus spelaeus*, and brown bear, *Ursus arctos*, lineages, *Current Biology*, **11** (2001), 200–203.

[54] E.M. Geigl, On the circumstances surrounding the preservation and analysis of very old DNA, *Archaeometry*, **44** (2002), 337–342.

[55] F.M. Guarino, F. Angelini, G. Odierna, M.R. Bianco, G. DiBernardo, A. Forte, A. Cascino, M. Cipollaro, Detection of DNA in ancient bones using histochemical methods, *Biotechnic and Histochemistry*, **75** (2000) 2, 110–117.

[56] E.M. Geigl, Inadequate use of molecular hybridisation to analyze DNA in Neandertal fossils, *American Journal of Human Genetics*, **68** (2001), 287–290.

[57] P. Shipman, G. Foster, M. Schoeninger, Burnt bones and teeth: an experimental study of color, morphology, crystal structure, and shrinkage, *Journal of Archaeological Science*, **11** (1984), 307–325.

[58] S. Parker, *An Experimental and Comparative Study of Cremation Techniques,* Unpublished M.Sc. dissertation, Department of Archaeology and Prehistory, University of Sheffield, 1985.

[59] N.P. Chandler, Cremated teeth, *Archaeology Today*, August (1987), 41–45.

[60] M.C. Stiner, S.L. Kuhn, S. Weiner, O. Bar-Yosef, Differential burning, recrystallization, and fragmentation of archaeological bone, *Journal of Archaeological Science*, **22** (1995), 223–237.

[61] Bennett J.L. Thermal alteration of buried bone, *J. Archaeol. Sci.*, **26** (1999), 1–8.

[62] K.A. Murray, J.C. Rose, The analysis of cremains: a case study involving the inappropriate disposal of mortuary remains, *Journal of Forensic Science*, **38** (1993) 1, 98–103.

[63] D.W. Owsley, Identification of the fragmentary, burned remains of two US journalists seven years after their disappearance in Guatemala, *Journal of Forensic Science*, **38** (1993) 6, 1372–1382.

[64] J.L. Holden, P.P. Phakey, J.G. Clement, Scanning electron microscope observations of heat-treated human bone, *Forensic Science International*, **74** (1995b), 29–45.

[65] K.A.R. Kennedy, The wrong urn: commingling of cremains in mortuary practices, *Journal of Forensic Science*, **41**(1996) 4, 689–692.

[66] C. Cattaneo, S. DiMartino, S. Scali, O.E. Craig, M. Grandi, R.J. Sokol, Determining the human origin of fragments of burnt bone: a comparative study of histological, immunological, and DNA techniques. *Forensic Science International*, **102** (1999), 181–191.

[67] H. Newesely, Chemical stability of hydroxyapatite under different conditions, in: G. Grupe, B. Herrmann (Eds.), *Trace Elements in Environmental History – Proceedings of the symposium held from June 24th to 26th, 1987, at Gottingen*, Springer-Verlag, Germany, pp. 1–16, 1988.

[68] A. Ravaglioli, A. Krajewski, G.C. Celotti, A. Piancastelli, B. Bacchini, L. Montanari, G. Zama, L. Piombi, Mineral evolution of bone, *Biomaterials*, **17** (1996) 16, 617–622.

[69] K.D. Rogers, P. Daniels, An X-ray diffraction study of the effects of heat treatment on bone mineral microstructure, *Biomaterials*, **23** (2002) 12, 2577–2585.

[70] H.Y. Juang, M.H. Hon, Effect of calcination on sintering of hydroxyapatite, *Biomaterials*, **17** (1996) 21, 2059–2064.

[71] J.L. Holden, J.G. Clement, P.P. Phakey, Age and temperature related changes to the ultrastructure and composition of human bone material, *Journal of Bone and Mineral Research*, **10** (1995a) 9, 1400–1409.

[72] T.J.U. Thompson, *A Preliminary Investigation Into the Influence of Burning on the Ability to Sex the Pelvis*, Unpublished MSc dissertation, Department of Archaeological Sciences, University of Bradford, 1999.

[73] M. Raspanti, S. Guizzardi, V. De Pasquale, D. Martini, A. Ruggeri, Ultrastructure of heat-deproteinated compact bone, *Biomaterials*, **15** (1994) 6, 433–437.

[74] G. Quatrehomme, M. Bolla, M. Muller, J.-P. Rocca, G. Grevin, P. Bailet, A. Ollier, Experimental single controlled study of burned bones: contribution of scanning electron microscopy, *Journal of Forensic Science*, **43** (1998) 2, 417–422.

## Chapter 4

# The Use of X-ray Scattering to Analyse Parchment Structure and Degradation

Craig J. Kennedy and Tim J. Wess

*Structural Biophysics Group, School of Optometry and Vision Science, Cardiff University, King Edward VII Avenue, Cardiff, CF10 3NB, UK*
*Email: KennedyC1@cardiff.ac.uk*
*Email: WessTJ@cardiff.ac.uk*

**Abstract**

Parchment is a collagen-based, historically important biomaterial that contains many layers of information, from text written on the surface, to the structure of the material itself. The degradation of historical parchments is often attributed to inappropriate storage conditions, although other factors may also accelerate the decay of collagen within the parchment, such as harsh cleaning or manufacturing techniques that involve extreme variations in pH or mechanical treatment.

X-ray diffraction at small and wide angles is an ideal tool for analysing the structure of the collagen, which gives the parchment its strength and durability over time. This chapter describes how small angle X-ray scattering (SAXS), in conjunction with sodium dodecyl sulphate polyacrylamide gel electrophoresis (SDS-PAGE) and thermal techniques, and wide angle X-ray diffraction (WAXD) can give a detailed account of the collagen structure within the parchment and indicate any deterioration brought about by laser-cleaning.

**Keywords:** X-ray diffraction, parchment, collagen, laser cleaning.

**Contents**

# 1. PARCHMENT

## 1.1. Parchment as a historical biomaterial

For millennia, parchment has been used as a writing medium. Perhaps the most famous example of historical parchment is the Dead Sea Scrolls, a group of documents which date from 300 BC to 70 AD, and were found in 1945 in a series of caves near the Wadi Qumran, approximately 2 km from the Dead Sea [1]. The religious texts displayed on the manuscripts have been of great importance to Middle Eastern historians. Notable examples of documents that have shaped the history of the world have been written on parchment: the Constitution of the United States of America (1787), which laid down the rights and responsibilities of the newly founded nation; the Treaty of Versailles (1919), which was the final peace agreement that ended the First World War; and the Treaty of Rome (1957), which saw the formation of the European Union. Two famous British examples of parchment are: the Doomsday Book and Magna Carta. The Doomsday book, commissioned in 1085 by William the Conqueror, contained the records for over 13 000 settlements in England at that time. The Magna Carta, authorised by King John of England in 1215, is considered as the corner stone of liberty and the chief defence against arbitrary and unjust rule in England. To this day, all Acts of Law that pass through the British Parliament are written on parchment.

Parchment is similar to leather in that both are biological materials processed from the skin of animals, usually cattle, sheep and goats [2]. Many of its structural features are derived from skin, and thus it is not a uniform structure in cross section, but is made up of distinct layers [3]. As a biomolecular composite, parchment is subject to deterioration due to the effects of atmospheric UV radiation, sulphur dioxide and microbial attack [4–6]. This realisation has led to an awareness that the historical written records, from the Dead Sea Scrolls to medieval European history, is under threat from increased pollution levels, damaging storage conditions, persistent humidity, and harsh methods of cleaning. To preserve the cultural heritage of nations and religions, a number of scientists have studied the manufacture and degradation of parchment with the aim of better understanding parchment breakdown. It is hoped that techniques to slow or prevent parchment deterioration, or to regenerate the parchment structure, can be developed [7–9].

In recent years, renewed attention has been given to the study and restoration of historical parchment. Large, multidisciplinary projects such as the EU projects Microanalysis of Parchment (MAP; Ref. [7]) and Improved Damage Assessment of Parchment (IDAP; www.idap-parchment.dk) have used X-ray diffraction methods in conjunction with other biochemical, mechanical, thermal and visual techniques to provide a broad overview of the characteristics of a large number of historical samples from a number of European countries. This chapter will focus on the analysis of parchment degradation through X-ray scattering techniques, in the contexts of damage assessment and conservation practice.

## 1.2. The structure of collagen within parchment

Parchment, following an extensive manufacturing procedure, is composed mostly of the protein collagen. As the loss of collagen structure is linked to the degradation of parchment over time, it is important to understand the structure of collagen that is present in parchment. By examining the collagen in parchment, from new through to the final degraded collagen of historic parchments, a clearer understanding of parchment degradation may be obtained.

In skin, and subsequently in parchment, collagen is the predominant protein present which provides mechanical strength [10]. At the ultrastructural level, collagen exists in the form of fibres. These fibres are composed of fibrils, which are made up of collagen molecules, which in turn are comprised of individual peptide chains, providing collagen with a discrete structural hierarchy.

Collagen fibres are approximately 50–300 μm in diameter, and are composed of tightly packed collagen fibrils [11]. The alignment of collagen fibres is an important factor in the overall mechanical characteristic of a tissue; in skin, the fibres are predominantly arranged in a two-dimensional felt-like network [12,13]. The fibres lie at random orientations in two dimensions over a large area of the skin, providing tensile strength in the plane of the parchment [14], although collagen fibres in skins taken from the spine or under the legs of an animal tend to display some preferential orientation.

Collagen fibrils are the principal, tensile strength-bearing components of connective tissues [15]. Of the twenty-nine known types of collagen molecules, types I, II, III, V and XI are capable of self-assembling to form fibrils [16]. Collagen fibrils are approximately cylindrical with diameters ranging between 10 and 500 nm [17], and range from 40 to100 nm in skin [18].

The axial direction of collagen fibrils exhibits a long-range order. The 300 nm-long collagen molecules are staggered relative to their neighbouring molecules by a regular distance, *d*, which is typically ~67 nm in tendon, or ~65.5 nm in skin [19,20], comprising gap and overlap regions (Fig. 1). The presence of the gap region is a consequence of staggering structures 300 nm in length at 67 nm intervals. The collagen molecule, at 300 nm

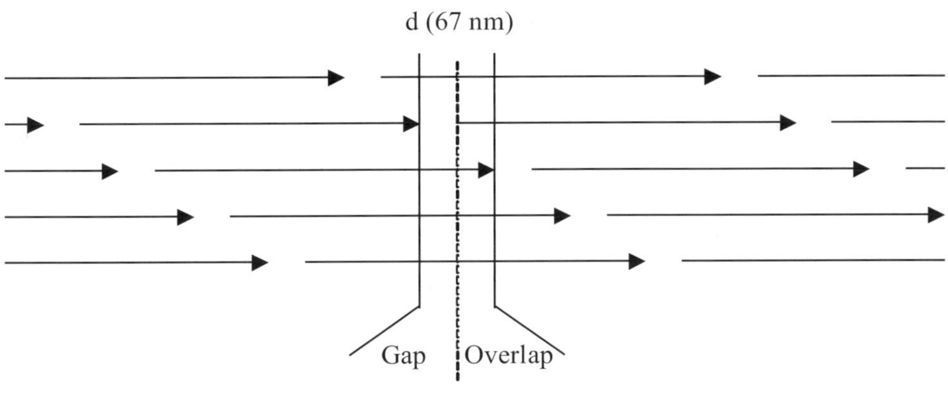


**Fig. 1.** The staggered array of collagen molecules, represented by arrows. Within each *d*-spacing are 4 complete molecular segments, and 1 half molecular segment. One *d* period is highlighted, with the gap and overlap regions labelled for clarity.

in length, consists of five segments: four of length *d* (67 nm) and one of length $d \times 0.46$. The arrangement of molecules of total length ~$4.4 \times d$ (300 nm), aligned in a parallel fashion and staggered by integral multiples of *d*, is known as the Hodge–Petruska model [21,22]. The *d* (~67 nm) repeating unit is characteristic of collagen. As the molecular length is not an exact multiple of the *d* period, the arrangement between linearly adjacent molecules results in a gap region. The gap region accounts for 0.54 of *d*, and the overlap subsequently comprises 0.46 of *d* [23]. The quarter-staggered array of collagen molecules allows the strength of the molecules to be translated to the next level of the structural hierarchy of collagen. This is essential to the ability of collagen to function within connective tissues. X-ray diffraction has shown that in dried collagen samples including parchment, the axial electron density profile of the collagen *d* period is altered, with the sharp interface between the gap and overlap regions becoming less apparent [24]. Additionally, the 65.5 nm stagger of the molecules in the skin is reduced to ~64 nm.

The collagen molecule comprises a triple-stranded rope-like structure formed by three interwound polypeptide chains, called α-chains. In skin, the main collagen types present are types I and III, in which the molecules are axially aligned to produce fibrils. The long middle section of the polypeptide chains, which exist in a triple-helical conformation, invariably has the amino acid sequence glycine-X-Y, where X and Y are any amino or imino acids, most commonly proline and hydroxyproline (Fig. 2), which are required for the formation of triple helix. Each molecule contains short regions at the N- and C-terminals that do not conform to this triplet repeat; these sections are termed telopeptides. The telopeptide regions contain lysine residues, which are implicated in covalent cross-links between neighbouring collagen molecules. The telopeptides and associated cross-links have a vital role in maintaining the structure of the collagen fibrils [23].

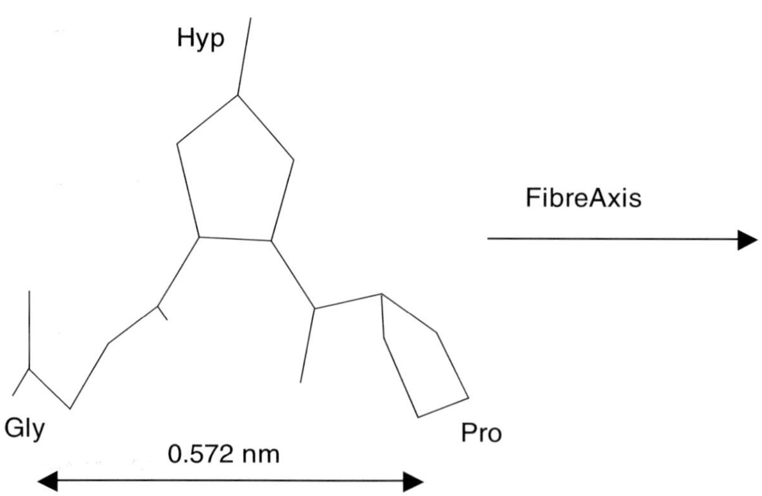


**Fig. 2.** Segment of a (Gly-Hyp-Pro) triple helix, indicating the relative sizes of the three amino acids. Atomic co-ordinates are from Bella *et al.* [25] PDB entry 1CAG.

### 1.3. Degradation of parchment

The processes of parchment manufacture and degradation alter the structure of the collagen, from an intact, fibrillar structure to a more disordered system. This constitutes the beginning of the collagen degradation pathway. Over time, this process is accelerated by long-term factors such as storage in hot or humid atmospheres, and short-term factors, such as fire, flood or harshly applied cleaning techniques.

Collagen degradation in parchment can be initiated by biological agents such as bacteria, fungi and rodents [5]. In humid environments at temperatures above 40°C, the prospect of microbial attack increases. Parchment has a pH of between six and eight, and is a source of nutrition for many microorganisms. There are three main degradation pathways for collagen: oxidation, hydrolysis and gelatinisation [26].

The oxidation of the collagen molecules can occur in the side chains of individual amino acid residues, the main chain of the collagen molecule, or between the amino group of an amino acid residue and its associated Cα-atom. Oxidation of the side chains is manifested as a reduction of the number of basic amino acids such as arginine, hydroxylysine and lysine, and an increase in the number of acidic amino acids such as glutamic acid and aspartic acid, which originate from glutamine and asparagine deamination (Fig. 3). The level of oxidation of collagen in parchment can be assessed by measuring the ratio of basic to acidic amino acids (B/A ratio). In fresh collagen, this ratio is 0.69, but as the collagen undergoes oxidative change, the ratio decreases. Historic parchments have shown B/A ratios as low as 0.5 [7,27]. Oxidation caused by free radicals is capable of breaking

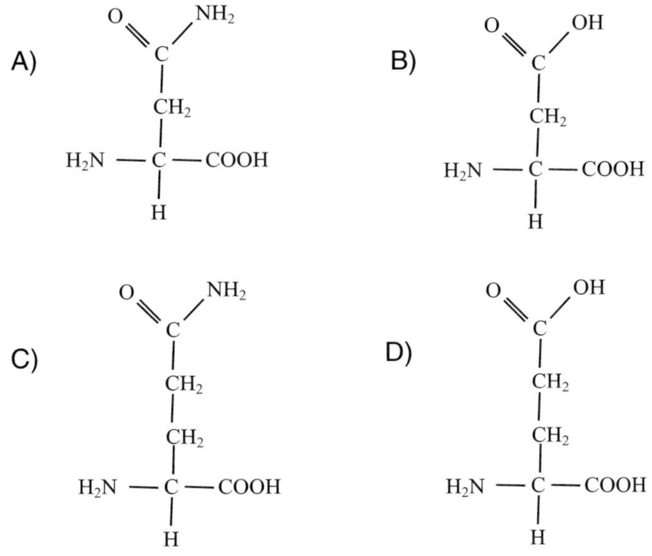


**Fig. 3.** The chemical structures of asparagine (A) and glutamine (C), and their deamination products aspartic acid (B) and glutamic acid (D).

the N–C covalent bonds that link neighbouring amino acid residues. The effect of this is cleavage of the polypeptide chains that comprise the collagen molecule. Oxidative cleavage of the collagen molecules occurs preferentially at tyrosyl residues on the collagen molecule [7,28] or in regions of charged residues [29].

Hydrolysis can be caused by acids, most commonly from combinations of water and atmospheric pollutants, such as $SO_2$ and water mixing to form sulphuric acid. Acids act in conjunction with water to bring about a cleavage in the main chain of the collagen molecule. The smaller peptides that result from this can undergo further hydrolysis; heavily deteriorated parchments may consist of smaller polypeptide chains compared to less degraded parchments. Through both oxidative [30] and hydrolytic [31] breakdown processes, the large 300 nm collagen molecules are broken into smaller fragments. This has an effect on the hierarchical structure of collagen, since the collagen molecules that have been cleaved no longer contribute to the strength of the collagen fibrils. Overall, this reduces the stability of the collagen hierarchy, and is the characteristic of collagen degradation in parchment.

A further mode of degradation of the collagen molecules in parchment is gelatinisation – the conversion from the fibrillar arrangement of molecules in a triple helix form to a random conformation [32]. There are a number of methods of inducing gelatinisation in collagen, including the addition of water and heat [7], or bathing in acidic or alkaline conditions [33].

Water competes with the existing hydrogen bonds within collagen and attempts to form new bonds with the molecule. This occurs when water is present in the system and hydrogen bonds are in a position within the molecule where they are open to attack from the water molecules. The action of heat makes water-induced gelatinisation more likely to occur [34]. As the heat increases, the hydrogen bonds gain mobility, enhancing the chance of interaction with water. When this occurs, the three chains of the molecule are no longer held together and are free to form individual, less-ordered structures.

During both acid and alkaline extraction of collagen, hydrolytic changes occur, leading to the release of collagenous material which is subsequently gelatinised at neutral pH at temperatures over 60°C [33]. For acid extraction of collagen, the tissue is typically soaked in dilute acid, followed by extraction with warm water at an acidic pH. The use of acid to produce gelatin from collagen is a harsh technique, with cleavage of the collagen molecules occurring in addition to unraveling of the triple helix [35]. The use of alkaline solutions to produce gelatin has been shown to act in a similar manner to acid solutions, but is more extreme in its damage to the collagen molecules. The appearance of additional N-terminal residues indicates that a significant number of peptide linkages are broken in the alkali pre-treatment [35,36]. The amide groups of glutamine and asparagine residues are released, as occurs during liming, resulting in a gelatin with an isoelectric point of ~pH 5 [37].

Gelatinisation is more likely to occur in partially degraded collagen molecules compared to native intact collagen, as the energy required to denature a shortened triple helix is lower than that of an intact one [38]. Even with damage to the collagen main chain, a factor may need to be introduced which would induce gelatinisation of the collagen, *e.g.* the addition of water. Collagen within parchment may exist in this "pre-gelatinous" condition for many years before gelatinisation is induced (R. Larsen, personal communication).

## 2. TECHNIQUES

### 2.1. X-ray diffraction as a tool to analyse parchment structure

Whilst many X-ray diffraction studies of native collagen (*e.g.* from rat tail tendon) have been conducted, relatively few have been carried out on parchment. The first major analysis came from wide angle X-ray diffraction (WAXD) of Dead Sea Scrolls samples by Weiner *et al.* [32]. This technique is capable of describing molecular-level details of samples, in the range of approximately 0.1–20 nm. In terms of collagen, this provides information regarding the molecule–molecule interactions within a fibril, and the helical characteristic of the collagen polypeptides (Fig. 4).

Weiner *et al.* used WAXD to determine the extent of degradation in parchment [32]. Two peaks from the WAXD profile were used: an equatorial peak at ~1 nm, present only in fibrillar collagen samples, which arises from the intermolecular interactions within a collagen fibril; and a peak at ~0.45 nm, attributed to amorphous polypeptide features, which

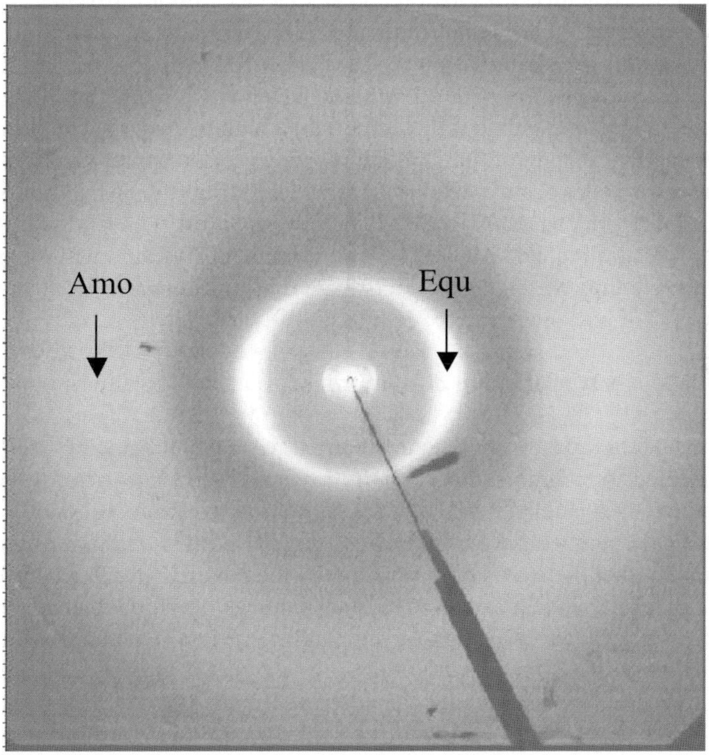


**Fig. 4.** Wide angle X-ray diffraction image of collagen within parchment taken at beamline ID18F at the ESRF, France. Observed are the equatorial reflection (Equ) due to molecular interactions within a fibril, which occurs at 0.85 $nm^{-1}$, and the reflection produced by amorphous interactions within a polypeptide (Amo), at 2 $nm^{-1}$.

is constant in fibrillar collagen and gelatin samples. The sum of the slopes of the equatorial peaks was divided by that from the amorphous peaks providing a collagen to gelatin (C:G) ratio. This allowed for a comparison between a number of important samples.

Following the work of Weiner *et al.*, a number of advances were made in X-ray diffraction technology [32]. For example, the photographic film has largely been replaced by charge-coupled devices (CCDs) as a means to collect diffraction images. This allows for easier computer-based analysis of the diffraction profiles, and more in-depth analysis of the features present in the images. Improvements have been made to the dynamic range and the ability to avoid high count-rate saturation of detectors. The level of background scattering can be better estimated from CCD images; photographic film is more prone to saturation, making this estimation extremely difficult. Additionally, CCDs instantly produce X-ray diffraction images without the need to develop film; this development has led to rapid output of data, and subsequently has made high-throughput X-ray diffraction experiments a reality.

Advances were also made in the field of photon production. In 1980, at the time of Weiner's work, the first second generation synchrotron radiation (SR) source was opened in Daresbury, UK. Synchroton radiation sources produce highly parallel, high brilliance X-ray beams, reducing experimental time and improving data quality. Since then, third-generation synchrotron radiation sources have been developed, such as the European Synchrotron Radiation Facility (ESRF), France, which came online in 1994. Third generation SR sources provide exceptionally high quality, high intensity X-ray beams, such that an X-ray diffraction image of parchment typically takes approximately one second, compared to five minutes from a second generation SR source, or five hours from a lab-based X-ray source such as the NanoSTAR facility at Cardiff University. In UK, the new third generation SR source, DIAMOND becomes online in 2007 and will provide advanced SR capabilities to cultural heritage artefacts from that time onwards.

## 2.2. Small angle X-ray scattering of parchment

Non-destructive techniques or techniques that use very small samples, *e.g.* on a micron scale, are preferred to techniques that require loss of sample integrity for the analysis of historical materials, especially if they can access similar information. Smaller sampling of valuable historical documents would allow assessment of the condition of the material as a whole, assuming that there was little variation in the material structure. Recently, micro-drills have been developed that can remove small samples from documents, which can then be examined to provide an indication of the condition of the document as a whole [39]. Of even more interest to conservators and archivists is the prospect of non-destructive analysis of parchment; techniques that can describe the state of degradation at a molecular or supra-molecular level without the requirement for cutting or drilling a sample. Small angle X-ray scattering (SAXS) is one such technique.

Up-to-date technologies were used to describe SAXS analysis of historical parchment [40]. X-ray diffraction at smaller angles ($<6°$ at $\lambda = 1.54$ nm) gives information concerning long-range order of structures; it is a process that is analogous to X-ray diffraction. However, this technique uses a much greater sample-to-detector distance that allows longer periodicities and particle sizes in the nanometre range to be investigated. Small angle X-ray scattering does not rely on crystallinity in the sample, and is one of the few techniques that gives reliable information regarding nanoscales within amorphous materials.

In the case of collagen, the axial staggering of molecules is observed by the meridional series of reflections at small angles (Fig. 5). The meridional series within historical parchment was the basis for the analysis of Wess *et al.*, who demonstrated that in degraded samples the level of disorder present was higher, and the meridional reflections became less pronounced or disappeared altogether [34]. Crystalline lipids within a sample were also observed, as the head-to-head distance of phospholipid bilayers falls into the SAXS region.

## 2.3. Biochemical and thermal analysis: correlation to SAXS

In order to determine the usefulness of SAXS as a non-destructive technique to analyse parchment structure, SAXS was coupled with several micro-destructive techniques – differential scanning calorimetry, shrinkage temperature analysis and SDS-polyacrylamide gel electrophoresis – to provide a useful means of examining the extent of degradation in historical parchments. Experimental details for these techniques are found in Refs. [41,42].

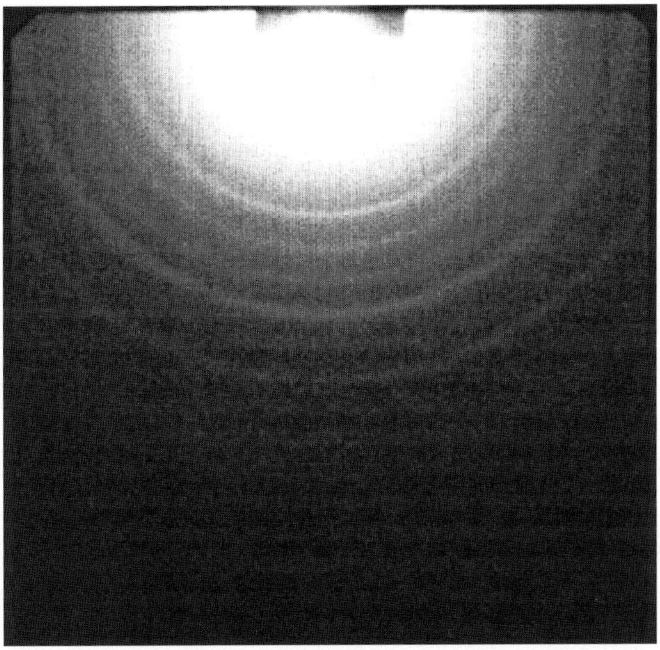

**Fig. 5.** A small angle scattering image of collagen from skin. The strong rings represent the meridional series of collagen, which is due to the axial electron density of collagen. The image was taken at station 2.1, SRS Daresbury, UK, at a sample to detector distance of 4 m (Image courtesy of C.A. Maxwell, Cardiff University).

Differential scanning calorimetry (DSC) is a method of measuring enthalpic changes in protein samples by controlled heating. The technique measures the degree of crystallinity and hydration in protein samples such as collagen [43]. The temperature of denaturation ($T_D$) is a direct measurement of the deterioration of a sample: samples with lower $T_D$ values are more disordered.

When collagen fibres are heated in water, they deform over a distinct temperature interval. This deformation is seen as shrinkage of the fibres and is registered in sub-intervals which rank the shrinkage intensity of the sample [7,44]. The shrinkage temperature ($T_S$) of collagen fibres can be a valuable indicator of their hydrothermal stability. Whilst this technique is closely related to DSC, the analysis conducted here is macroscopic in nature, and therefore provides information regarding the parchment structure on a larger scale than that observable by X-ray diffraction, DSC or SDS-PAGE.

SDS-polyacrylamide gel electrophoresis (SDS-PAGE) is employed to assess the molecular integrity of collagen within the parchment. Following cyanogen bromide (CNBr) digestion, which specifically cleaves proteins at methionine residues, fully intact collagen yields a specific banding pattern once subjected to SDS-PAGE [45]. However, should the collagen molecules have previously been damaged by, for instance, oxidative or hydrolytic cleavage of the main chains at random locations, then the banding pattern would appear less intense relative to the background, due to a spectrum of peptide lengths, on a gel.

SAXS data were reduced to one-dimensional linear profiles in accordance with Ref. [34]. A parameter of sample ordering was measured by SAXS; from this, a ratio of crystallinity was devised. This involved taking the integral of the area of a linear profile corresponding from the 6th to the 9th orders of diffraction ($I_1$) and dividing that integral by the integral of the entire linear trace ($I_2$; Fig. 6). This parameter is based upon the assumption that the meridional orders of collagen represent the crystalline component of the collagen fibrils,

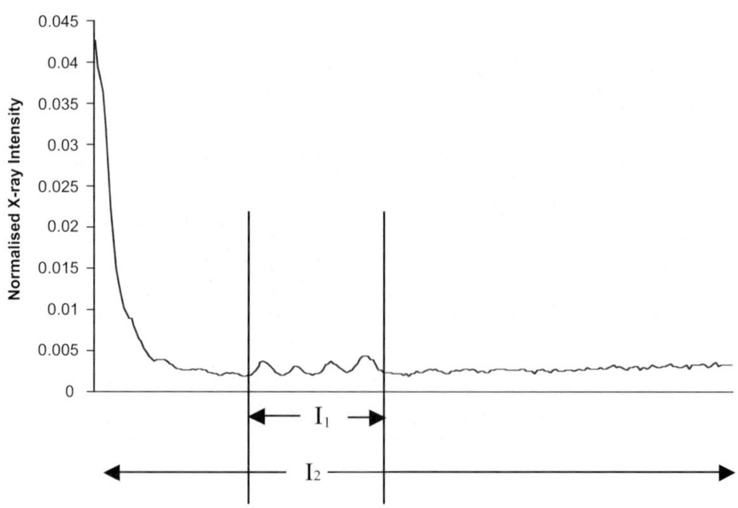


**Fig. 6.** Linear profile from parchment sample 01. $I_1$ denotes the area where the integral of the intensity is taken between the 6th and the 9th orders of collagen. $I_2$ denotes the area of the integral of the entire linear profile.

whilst the overall scattering is due to the total X-ray scattering from crystalline and non-crystalline regions. In this manner the crystallinity is measured in biological polymer systems such as cellulose and silk [46]. The linear profiles are first treated by removing the area near the centre of the diffraction image to remove any effects generated by the beam-stop. The minimum values were also subtracted from the profiles to remove any underlying background from the sample that had not been removed from the diffraction image. Finally, the profiles are normalised by their integrated intensities to account for sample density.

Four historical and two new samples were selected from a larger collection of parchment samples for complementary analysis by the methods discussed above. These samples represented a great deal of variation in physical features, such as thickness and colour, as well as in SDS-PAGE and SAXS profiles. As all the samples were provided by the same source (National Archives of Scotland) and are of approximately the same age, the analysis of these samples is assumed to be representative of the larger set.

## 3. RESULTS

The ratio of crystallinity, denaturation temperatures and shrinkage temperatures of the sample are shown in Table 1. The crystallinity ratios range from ~0.05 to ~0.2, indicating that the integrity of the collagen structure varies across the sample range. The denaturation temperatures vary between 46 and 64°C. The range of temperatures of denaturation displayed by the samples suggests that the parchments have undergone varying levels of deterioration; as the $T_D$ of a sample decreases, its thermal stability is lowered, indicating that the collagen has undergone damage which renders it susceptible to denaturation at low temperatures. Mean values for $T_S$ measurements are shown in Table 1. These values are consistent for repeat experiments, within a margin of ±2°C. Samples 01 and 03 show higher values for $T_S$ than samples 02 and 04, indicating that these two samples have a higher degree of hydrothermal stability at the macroscopic level.

The main features observed in the SDS-PAGE profiles were sharp peaks relating to the positions of the α2-CB3,5 peptide which represents the major triple helical constituent of

**Table 1.** Values for the ratio and crystallinity (±0.008), denaturation temperature (±1.6°C) and shrinkage temperature (±2°C) of the samples analysed. $T_D$ and $T_S$ values for the new parchment samples were obtained from the literature

| Sample name | Ratio of crystallinity | $T_D$ (°C) | $T_S$ (°C) |
|---|---|---|---|
| New Calf | 0.199 | 51.5–64* | 44–55# |
| New Goat | 0.196 | 51.5–64* | 44–55# |
| 1 | 0.093 | 62.3 | 55 |
| 2 | 0.05 | 63.04 | 45 |
| 3 | 0.068 | 57.44 | 53 |
| 4 | 0.133 | 46.68 | 44 |

* Reference [47].

\# Reference [47].

the α2 chain of type I collagen, and a broad peak that corresponds to both the α1-CB7 and α1-CB8 peptides [48]. Samples that display these peaks have intact collagen molecules prior to CNBr treatment; samples that do not display these peaks underwent some molecular scission beforehand.

The experiments conducted have yielded a range of results for the samples examined here. Historical samples were provided by the National Archives of Scotland, and were all dated from the eighteenth and nineteenth centuries. Of the historical samples, 01 (dated 1765) and 04 (dated 1827) yielded the highest ratios of crystallinity, suggesting that the molecular arrangement of collagen molecules within the fibrils remains relatively intact. Samples 02 and 03 (dated 1769 and 1775, respectively), on the other hand, show reduced ratios of crystallinity, suggesting that the collagen molecules are no longer arranged in a fibrillar formation. The reason for this variation is not clear. Sample age does not appear to be the governing factor, with the oldest and youngest samples displaying the highest ratios. Other possible reasons for the degradation of the samples could include the condition of the animal the skin was taken from, the parchment preparation technique, storage conditions over the centuries and effects of short-term damage events such as fire or flood.

DSC shows a range of temperatures at which collagen denatures ($T_D$). It has been shown [47] that new parchment samples display $T_D$ values between 51.5 and 64°C, depending on preparation technique. Samples 01 and 02, nearly 250 years old, display $T_D$ values that are of a similar level to high-quality new parchment. There are two possible explanations for these high values. First, the parchment samples had high $T_D$ values when new, and underwent very little deterioration over time; this is impossible to assess as no known measurements were taken from these samples at the time of manufacture. The second possibility is that additional cross-links have been introduced into the collagen which stabilise the collagen molecules and increase their thermal denaturation temperature. This may be possible as remains of parchment and historical documents indicate that in some cases parchment was treated with vegetable tannins to colour or tan it [49]. Samples 03 and 04 displayed lower values of $T_D$, indicating a loss of hydrothermal stability in their molecular and fibrillar structure. This indicates that these samples have endured conditions that are more damaging to their structure than the better preserved samples.

Samples displayed $T_S$ measurements in the range of 44–55°C. New parchments display $T_S$ values of between 52 and 62°C [7]; samples 01 and 03 display $T_S$ values within this range. As with the other techniques employed, $T_S$ measurements show variations between samples; 02 and 04 display low $T_S$ values, indicating reduced fibre stability.

Major variations in the SDS-PAGE analysis of digested peptides concerned the presence and position of three peaks that represent the α2-CB3,5, the major triple helical constituent of the α2 chain of type I collagen, α1-CB7 and α1-CB8 peptide digests. In terms of the three main peaks observed in the gels, sample 01 displayed strong peaks for these digests; sample 02 showed little or no peaks. Sample 03 showed some variations in this regard, with some samples from this document displaying weak peaks and some samples displaying no peaks. This indicates that variation exists within the sample in terms of the molecular integrity of the collagen, depending on the area from which the sample was taken. Sample 04 displayed a weak banding pattern in almost all cases, indicating that the collagen molecules were badly damaged before CNBr treatment.

### 3.1. Comparative analysis of results

Results from the samples analysed here indicate variation in terms of fibrillar crystallinity, macroscopic and molecular denaturation temperatures, and molecular integrity. When only samples 01, 02 and 03 are compared, there appears to be a reasonable correlation across the range of experimental results. For instance, sample 01 displays a high ratio of crystallinity, high $T_D$ and $T_S$ temperatures and a strong SDS-PAGE banding pattern, whilst sample 02 displays a low ratio of crystallinity, low $T_S$ values and a weak SDS-PAGE banding pattern. Sample 03 appears to be intermediate in all of these regards. Sample 04, however, is inconsistent across the range of experimental procedures carried out here; it has the highest ratio of crystallinity, the lowest $T_S$ and $T_D$ values and a very weak SDS-PAGE banding pattern.

A number of possibilities are presented by these results. The first is that there are quite extreme variations in the collagen structure within a sample. However, within each experimental procedure, the samples display consistent results, even when a number of samples are taken from a range of areas of each document, making this explanation unlikely. The second possibility is that the samples may be degraded at one level of the collagen structural hierarchy, but this may not manifest itself in other levels of the hierarchy. This would partially explain some of the results observed here. For example, sample 04 displays low temperatures of molecular denaturation and fibre shrinkage, and a weak SDS-PAGE banding pattern. These results would indicate that the collagen molecules and collagen fibres are damaged. However, SAXS indicates that the molecules within this sample are aligned in a fibrillar formation. This suggests that this sample may exist in a pre-gelatinous state. In this instance, the collagen molecules are damaged still retaining their molecular arrangement. Only in the presence of an external factor does this arrangement change; in these cases, the addition of heat and water for thermal measurements, and CNBr for SDS-PAGE will cause the damaged molecules to change to a gelatinous condition. As SAXS is a non-destructive technique that does not act upon the molecular arrangement of the molecules, or introduce a force that will cause the molecules to change conformation, it cannot measure the pre-gelatinous state.

## 4. SURFACE TO SURFACE ANALYSIS OF PARCHMENT CROSS SECTIONS

X-ray diffraction is an analytical method that can be used to detect changes in the structure of collagen within the parchment, which may occur in decay processes or through applied conservation or cleaning treatments. The advent of microfocus X-ray technology has led to the development of intense micron-sized X-ray beams [50], which can be used for X-ray microdiffraction experiments where small areas of textural variation can be probed in a sample. This technology also allows simultaneous measurements of fluorescence emission, allowing elemental mapping of small regions of a sample. Microfocus X-ray diffraction can be used to make a detailed map of nanotexture within a biologically based material, such as bone [51]. However, the technique can potentially be applied to all materials containing nanostructural variation on a microscopic length scale. This technique has been applied to a number of collagenous tissues, such as teleost scales [52], cornea [53] and osteonic lammelae [54].

Kennedy *et al.* [55] used microfocus X-ray beams to analyse cross sections of parchment samples to gain a clearer understanding of the parchment structure in cross section. At beamline ID18F at the European Synchrotron Radiation Faclity (ESRF), using a beam of 1.5 × 15 μm, up to 200 X-ray diffraction images could be taken from a 300-μm thick sample. Figure 7 displays a schematic diagram of beamline ID18F.

X-ray microdiffraction allows for analysis of features present only in specific areas of the parchment, such as at the surface. Kennedy *et al.* [55] utilised this technique to analyse both the collagenous and the non-collagenous properties of parchment. Twelve historical parchment samples were analysed, dating from the eighteenth and nineteenth centuries, provided by the Conservation Workshop, National Archives of Scotland, UK.

Structural features of parchment were investigated by microfocus X-ray diffraction, including variations in the orientation of collagen fibrils relative to the plane of the parchment, identification and location of mineral phases present in cross sections, and the state and presence of crystalline lipids. X-ray fluorescence profiles, obtained simultaneously with the X-ray diffraction patterns, provide a description of the elemental components of the cross sections, which complements the diffraction data to create a more complete account of the characteristics that comprise the parchment structure.

The alignment of collagen fibrils within parchment samples were believed to be roughly in the plane of the parchment surface, although no quantification of this has been conducted.

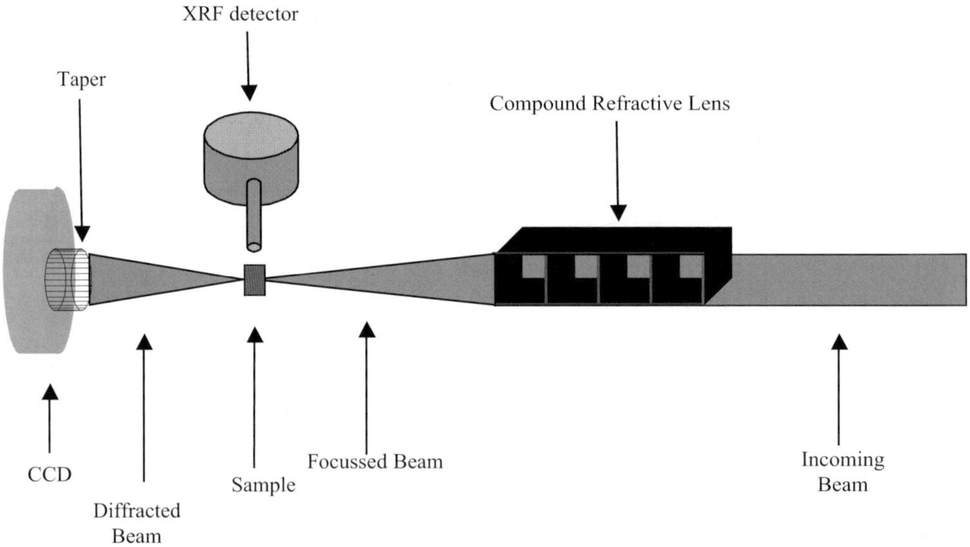


**Fig. 7.** Schematic layout of beamline ID18F at the ESRF. The X-ray beam is focussed to 1.5 × 15 μm by the compound refractive lens. The CCD collects the diffraction images, whilst the Si(Li) detector collects the fluorescence profiles.

Surface-to-surface scans of parchment have allowed for this to be measured and quantified. The radial width of the equatorial collagen diffraction peak at 0.85 $nm^{-1}$, which is attributed to interactions between neighbouring collagen molecules, was used as an indicator of collagen alignment. If the collagen fibrils were not well aligned, this peak would appear broad; consequently, if the fibrils are well aligned, this peak would appear to be sharp. Figure 8 displays how the full width half maximum (FWHM) values are calculated; in all cases the FWHM is approximately 80°. By performing this analysis on all images taken from a parchment cross section, orientation maps can be constructed that gives a clear indication of the variations in collagen orientation in parchment.

### 4.1. Non-collagenous components in parchment cross sections

Non-collagenous components of parchment observed by X-ray microdiffraction in cross section include lipids and minerals. The source of the lipids, shown by a characteristic diffraction peak at ~4.6 nm, is not fully understood. There are two main possibilities – either the lipids have survived through the parchment making process, or have been left as a result of microbial attack. The lipid diffraction peak appears equatorial to the main collagen fibril axis, suggesting that the lipids are packed between collagen fibrils. This would be consistent with lipids persisting from the original animal skin. In many cases, the lipids appeared to be very crystalline, showing several orders of diffraction. Through a

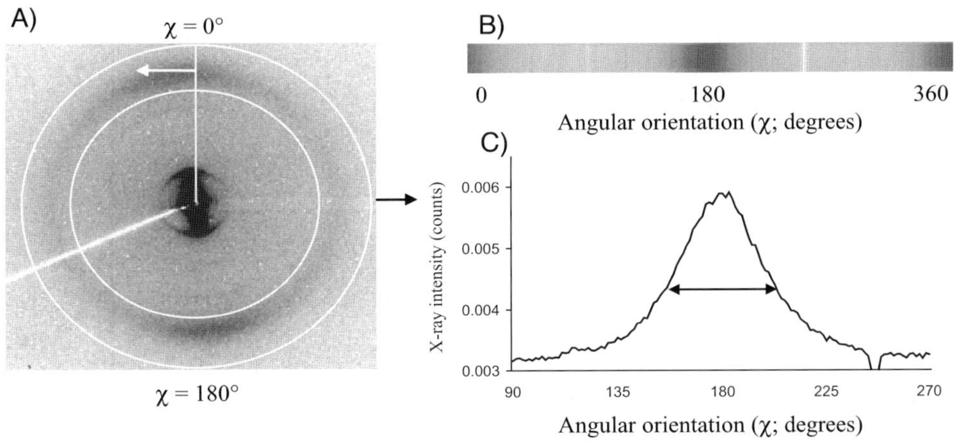


**Fig. 8.** Integration of area of the X-ray diffraction image of parchment sample 04 (A), which encompasses the 0.85 $nm^{-1}$ equatorial reflection. Two distances in $\theta$ are chosen that lie on either side of the 0.85 $nm^{-1}$ reflection. Integration then occurs radially between the two distances, *i.e.* 0.7 and 1.0 $nm^{-1}$, represented as white circles on the diffraction image, producing a linear map of angular intensity distribution. The intensity distribution map can be represented as a grey-scale image (B), where darker areas correspond to increased intensity; or as a graph (C) which allows for calculation of the full width half maximum (FWHM; shown by the arrow) of the peaks.

parchment cross section, the *d* spacing of the lipids varies between 4.4 and 4.6 nm, suggesting variations in the hydration state or biochemical composition of the lipid.

The presence of minerals in the parchment samples is evident by sharp peaks at very high angles of diffraction. The minerals are most likely to occur from parchment manufacture, where the skins are bathed in lime ($Ca(OH)_2$) to facilitate hair removal, and occasionally treated with chalk to alter the surface attributes of the parchment. Following drying, calcium carbonate ($CaCO_3$) is formed from the lime in parchment reacting with carbon dioxide ($CO_2$) from the air. In one sample, the minerals appeared to be present at the parchment surface, indicating the result of a chalk treatment. For all other samples which displayed mineral diffraction peaks, the peaks appeared at random throughout the cross section. This suggests that calcite may have entered the skin during the liming process. Mineral phases were identified using PCSIWIN software and the International Centre for Diffraction Data (ICDD) PDF-2 database. It was found that the minerals present were polymorphs of calcite such as aragonite and vaterite.

### 4.2. Microfocus X-ray fluorescence

Beamline ID18F at the ESRF is specially designed to allow for simultaneous X-ray diffraction and fluorescence experiments. Hence, for each X-ray diffraction image obtained, there is a corresponding fluorescence profile. The fluorescence detector is placed perpendicular to the incident X-ray beam in line with the sample and within the storage ring plane. The fluorescence detector is a Si(Li) detector (GRESHAM) with an active area of 30 $mm^2$ and 3.5 mm thickness with an 8 μm beryllium window used to detect the characteristic X-ray intensities generated by the excited area of the sample.

A CANBERRA 9660 digital signal processor and a CANBERRA 556A acquisition interface module (AIM) are used to collect the X-ray spectra [56]. X-ray fluorescence data is then treated using the computer program AXIL, which models a background prior to running a peak-fitting algorithm [57]. Following background subtraction, the values for peak heights and integrated intensities for a number of elements are obtained. The spectra are then corrected for the sensitivity of each element, taking into account the excitable cross-sectional area of each element, the fluorescence yield and absorption by the detector window.

As it appeared in most samples that the level of calcites present was not uniform throughout the parchment sections, or confined to the surfaces, X-ray fluorescence profiles were analysed to observe the levels of calcium through the sections. The X-ray fluorescence profiles confirm that the level of calcium through a cross section is variable, suggesting that pockets of low fibre density may house calcites.

## 5. LASER CLEANED PARCHMENT

The preservation and restoration of parchment has become a field of study for the scientific community [7,9]. As parchment ages, it can collect surface deposits that darken the parchment and make it harder to read clearly, and the presence of dirt is implicated as a factor that may accelerate the deterioration process [4]. Cleaning of parchment is an essential aspect of bringing historical documents to prominence in the modern day, allowing for the display of important artefacts of cultural heritage in a manner that is accessible to historians and non-historians alike [58]. Whilst cleaning in the conventional sense, using chemicals such as isopropanol, or techniques that involve water, cycling relative humidities, stretching and wetting parchments continue to be used in a conservation context, novel cleaning methods that can remove dirt from the parchment surface but not damage the integrity of the collagenous structure that maintains parchment as a durable biomaterial has advanced. An understanding of the impact of novel cleaning methods on the structure of collagen within parchment is essential if these techniques are to become widely available to the conservation community.

A novel cleaning method that has been developed is the laser cleaning of parchments. Laser cleaning has been used extensively in the cleaning of sculptures and buildings [59]. As parchment is a material comprised of collagen, which has a discrete structural hierarchy where its implicit strength is based on specific molecular and fibrillar interactions, it is much more susceptible to damage by lasers than materials such as stone or marble. An appropriate wavelength and energy level of the laser must be selected to ensure that no damage occurs in the collagenous structure of the parchment. Laser cleaning technology may have some advantages over conventional conservation methods: for example, laser cleaning is contactless, it is not overtly mechanical, and it does not require the use of chemicals that may disrupt the collagen structure. The impact of laser cleaning on aspects of parchment structure has previously been investigated by methods such as electron microscopy, infrared spectroscopy, amino acid analysis and thermal measurements [60–64].

### 5.1. Sample preparation: laser cleaning

Samples of new parchment were laser cleaned at the National Museums Liverpool [42] using a Q-switched Nd:YAG laser (Lynton Lasers Ltd.). Half of the samples were dirtied artificially using standard test dust before cleaning, and half were left undirtied. Parchments were then subjected to laser cleaning at three different wavelengths: 1064 nm (infrared), 532 nm (green) and 266 nm (ultraviolet). The pulse length was 5–10 ns.

For cleaning at 1064 and 532 nm, the parchment samples were mounted on a moveable computer controlled X–Y table and irradiated from above; the laser beam was delivered from a fixed position through an articulated arm with the end part of the hand piece including a focal lens. The parchment sample was scanned beneath the laser beam at a fixed rate, so that each part of the irradiated surface received approximately 22 pulses. The repetition rate was 1.25 Hz. For cleaning at 266 nm, the beam was delivered horizontally and directly from a bench top assembly connected to the laser system by the articulated arm.

### 5.2. SAXS of laser cleaned samples

As discussed previously in this chapter, X-ray scattering at small angles provides information regarding large-scale structures. At beamline 2.1, SRS Daresbury (UK), structures in the range of 0.01–0.15 $nm^{-1}$ can be observed using an 8.25 m camera [65]. In terms of collagen, this equates to the lower orders of the meridional series. The meridional series of collagen is produced by the characteristic gap and overlap regions in the axial direction of collagen fibrils. Changes in the relative intensities or positions of these reflections can be brought about by dehydration, heating or any other treatment that can alter the fibrillar structure of the collagen. If laser cleaning were to induce fibrillar damage to the collagen structure, SAXS can be used to discern the nature of such a change. In addition, large dirt particles that could be removed by laser cleaning can be characterised using SAXS.

Kennedy *et al.* [42] demonstrated that following laser cleaning at infrared (1064 nm) or green (532 nm) wavelengths, no changes in the structure of the collagen were detected by SAXS. However, after cleaning at the ultraviolet wavelength (266 nm) the first order of collagen diffraction appeared split, indicating some low coherence damage to the fibrillar structure of the collagen.The effect of laser cleaning on the dirt removed from parchment was also investigated. Artificial test dirt was used for these experiments. The test dirt was analysed using SAXS, applied to the parchment substrate, and collected on to a sheet of mica as it was removed by laser cleaning. Following collection on to mica, the dirt was analysed again using SAXS to observe any differences that may have been brought about by laser cleaning.

Guinier's law, which states that at very small scattering angles the angular dependence is a universal function of size [66,67], is capable of giving the radius of gyration $R_g$, in effect a measure of the particle size, which can be defined as the radius from a given axis at which the mass of a body could be concentrated without altering the rotational inertia of the body about that axis. The radius of gyration of the particles examined increases after laser cleaning, from 16.6 nm in the original artificial test dirt, to 23.4 nm after laser cleaning at 1064 nm and 23.1 nm after laser cleaning at 532 nm. Possible explanations for this are that either particles become joined together in some way during laser cleaning, or larger particles are preferentially removed, which is in agreement with Zheng *et al.* [68]. Also, at small angles, a distinct diffraction feature is observed corresponding to a spacing of 27 nm, which disappears after laser cleaning.

### 5.3. Microfocus X-ray diffraction of laser cleaned samples

SAXS characterisation of parchment usually entails the X-rays being transmitted throughout the entire depth of a sample. It is possible that cleaning a parchment sample could damage the collagen near the parchment surface, and leave the remaining collagen intact. In this case, SAXS would not necessarily detect the damaged collagen. To assess this possibility, microfocus X-ray diffraction [55] was used on laser cleaned parchment samples.

One parchment sample was cut into sections; one was used as a reference, one was laser cleaned at a wavelength of 1064 nm at a fluence level of 0.32 $J/cm^2$, and one sample cleaned for an extended period of time to induce damage to the collagen structure. Collagen to gelatin (C:G) ratios were calculated by updating the method of Weiner *et al.* [32]. To quantify the 0.85 and 2 $nm^{-1}$ reflections in the samples, the one-dimensional peak-fitting program XFIT (Collaborative Computational Project 13 (CCP13)) was used.

From linear profiles of the X-ray diffraction data, normalised by their integrated intensities, a polynomial background was removed, and the reflections were represented by Lorentzian distributions. The peaks are assumed to be Lorentzian in nature. Once the peaks were selected, a linearised least-squares algorithm was used to optimise the fit between the selected Lorentzian functions and the original linear profiles. The integrated intensity of the 0.85 $nm^{-1}$ peak is then divided by the integrated intensity of the 2 $nm^{-1}$ peak, producing a value for the relative amounts of collagen and gelatin (C:G). Values of 0 or less indicate that there is no 0.85 $nm^{-1}$ peak present, and hence the collagen is completely gelatinised.

Sample 04 was used in the laser cleaning analysis, as it has shown greater susceptibility to damage than other samples; a section of this sample was laser cleaned at a wavelength of 1064 nm at a fluence level of 0.32 $J/cm^2$. New parchment samples were used in the laser damaging part of the experiment. Microfocus X-ray diffraction analysis was conducted on the laser cleaned section and a reference section taken 2 mm from the site of laser cleaning. Laser cleaning appears not to induce gelatinisation of the collagen at the parchment surface. Intentional damage to the parchment surface with a laser, however, brings about a loss of intensity of the 0.85 $nm^{-1}$ reflection at the surface of the parchment that indicates a degree of gelatinisation.

## 6. CONCLUSIONS

X-ray diffraction is a tool capable of providing a detailed insight into the structure of parchment without the need for harsh sample preparation or destruction. Recent studies have allowed for characterisation of the collagen in parchment, from the deterioration in historical samples to the effects of laser cleaning. This chapter has provided a brief overview of the analysis of parchments by X-ray diffraction, a process that is still ongoing. The use of synchrotron radiation to analyse materials of high cultural heritage value is increasing, as demonstrated by the formation of the Synchrotron Radiation in Art and Archaeology (SR2A) group and the increasing number of publications in this field of study.

## REFERENCES

[1] D. Burton, J.B. Poole, R. Reed, A new approach to the dating of the Dead Sea Scrolls, *Nature*, **184** (1959), 533–534.

[2] R. Reed, *The Nature and Making of Parchment*, Elmete Press, Leeds, England, 1975.

[3] C.V. Horie (Ed.), Deterioration of skin in museum collections, *Polymer Degradation and Stability*, **29** (1990), 109–133.

[4] D.J. Bowden, P. Brimblecombe, Sulphur inclusions within parchment and leather exposed to sulphur dioxide, in: R. Larsen (Ed.), *Microanalysis of Parchment*, Archetype Publications, pp. 45–51, 2002.

[5] A.B. Strzelczyk, J. Karbowska, The role of microorganisms in the decay of parchment, *Acta Microbiologica Polonica*, **43** (1994), 165–174.

[6] M.N. Poglazova, Y.P. Petushkova, T.P. Svetlichnaya, Light optical and electron microscope study of microflora in ancient greek manuscript parchment, *Mikrobiologiia*, **57** (1988), 543–549.

[7] R. Larsen (Ed.), *Microanalysis of Parchment*, Archetype Publications, London, 2002.

[8] D. Fessas, A. Schiraldi, R. Tenni, L. Vitellaro Zuccarello, A. Bairati, A. Facchini, Calorimetric, biochemical and morpholigical investigations to validate a restoration method of fire injured ancient parchment, *Thermochimica Acta*, **348** (2000), 129–137.

[9] D.V. Parry, S.D. Ricks (Eds.), *Current Research and Technological Developments on the Dead Sea Scrolls*, E.J. Brill, Leiden, 1996.
[10] C.J. Kennedy, T.J. Wess, The structure of collagen within parchment: A review, *Restaurator*, **24** (2003), 61–80.
[11] S.L.-Y. Woo, G.A. Johnson, B.A. Smith, Mathematical modeling of ligaments and tendons, *Journal of Biomedical Engineering*, **115** (1993), 468–473.
[12] P.L. Kronick, P.R. Buechler, Fiber orientation in calfskin by laser light scattering or X-ray diffraction and quantitative relation to mechanical properties, *Journal of the American Leather Chemists Association* **81** (1986a), 221–231.
[13] P.L. Kronick, P.R. Buechler, Effects of beaming and tanning on collagen stability, studies by differential scanning calorimetry, *Journal of The American Leather Chemists Association*, **81** (1986b), 213–220.
[14] E. Hansen, F.S.N. Lee, H. Sobel, The effects of relative humidity on some physical properties of modern vellum, *Journal of the American Institute for Conservation*, **31** (1992), 325–342.
[15] P. Fratzl, K. Misof, I. Zizak, G. Rapp, H. Amenitsch, S. Bernstorff, Fibrillar structure and mechanical properties of collagen, *Journal of Structural Biology*, **122** (1997), 119–122.
[16] D.J.S. Hulmes, The collagen superfamily – diverse structures and assemblies, *Essays in Biochemistry*, **27** (1982), 49–67.
[17] D.J.S. Hulmes, T.J. Wess, D.J. Prockop, P. Fratzl, Radial packing, order and disorder in collagen fibrils, *Biophysical Journal*, **68** (1995), 1661–1670
[18] C.M. Kielty, I. Hopkinson, M.E. Grant, Collagen, The collagen family: Structure, assembly and organisation in the extracellular matrix, in: P.M. Royce, B. Steinmann (Eds.), *Connective Tissue and Its Heritable Disorders*, Wiley-Liss, New York, pp. 103–147, 1993.
[19] R.H. Stinson, P.R. Sweeny, Skin collagen has an unusual d-spacing, *Biochimica et Biophysica Acta*, **621** (1980), 158–161.
[20] B. Brodsky, E.F. Eikenberry, K. Cassidy, An unusual collagen periodicity in skin, *Biochimica et Biophysica Acta*, **621** (1980), 162–166.
[21] A.J. Hodge, J.A. Petruska, *Electron Microscopy vol. 1, Paper QQ-1*, in: S.S. Breese Jr. (Ed.), Academic Press, New York, 1962.
[22] A.J. Hodge, J.A. Petruska, Recent studies with the electron microscope on ordered aggregates of the tropocollagen molecule, in: G.N. Ramachandran (Ed.), *Aspects of Protein Structure*, Academic Press, New York, pp. 289–300, 1963.
[23] J.P. Orgel, T.J. Wess, A. Miller, The *in situ* conformation and axial location of the intermolecular cross-linked non-helical telopeptides of type I collagen, *Structure*, **8** (2000), 137–142.
[24] T.J. Wess, J.P. Orgel, Changes in collagen structure: drying, dehydrothermal treatment and relation to long term deterioration, *Thermochimica Acta*, **36** (2000), 119–128.
[25] J. Bella, M. Eaton, B. Brodsky, H.M. Berman, Crystal and molecular structure of a collagen-like peptide at 1.9Å resolution, *Science*, **266** (1994), 75–81.
[26] M. Derrick, Evaluation of the state of degradation of Dead Sea Scroll samples using FT-IR spectroscopy, *American Institute for Conservation Book and Paper Group*, **10** (1991), 49–65.
[27] R. Larsen, V. Barkholt, K. Neilsen, Amino acid analysis of leather. Preliminary studies in deterioration, accelerated ageing and conservation of vegetable tanned leather, *Das Leder*, **40** (1989), 153–158.
[28] C.L. Deasy, S.C. Michele Sr., A study of the oxidative degradation of gelatin and collagen by aqueous hydrogen peroxide solutions, *Journal of the American Leather Chemists Association*, **60** (1965), 665–674.
[29] R. Larsen, The possible link between the collagen sequence and structure and its oxidative deterioration pattern, in: R. Larsen (Ed.), *STEP Leather Project*, Research Report no. 1, The Royal Danish Academy of Fine Arts, pp. 59–74, 1994.
[30] C.L. Deasy, Degradation of collagen by metal ion-hydrogen peroxide systems I. Evidence for a free radical catalysed depolymerisation mechanism, *Journal of the American Leather Chemists Association*, **62** (1967), 258–269.
[31] J.H. Bowes, A.S. Raistrick, The action of heat and moisture on leather part IV: degradation of the collagen, *Journal of the American Leather Chemists Association*, **62** (1967), 240–257.
[32] S. Weiner, Z. Kustanovich, E. Gil-Av, W. Traub, Dead Sea Scroll parchments: unfolding of the collagen molecules and racemisation of aspartic acid, *Nature*, **287** (1980), 820–823.
[33] W.F. Harrington, P.H. von Hippel, The structure of collagen and gelatin, *Advances in Protein Chemistry*, **16** (1961), 75–138.

[34] B. Hassel, Heat-damaged parchment, *PapierRestaurierung*, **3** (2002) 4, 31–38.
[35] A. Courts, Structural changes in collagen. The action of alkalis and acids in the conversion of collagen into eucollagen, *Biochemical Journal*, **74** (1960), 238–247.
[36] C.L. Deasy, Determination of N-terminal amino acids of collagen by the 2,4-dinitrofluorobenzene method, *Journal of the American Leather Chemists Association*, **53** (1958), 203–206.
[37] A.W. Kenchington, A.G. Ward, The titration curve of gelatin, *Biochemical Journal*, **58** (1954), 202–207.
[38] R.A. Condell, N. Sakai, R.T. Mercado, E. Larenas, Quantitation of collagen fragments and gelatin by deconvolution of polarimetry denaturation curves, *Collagen Related Research*, **8** (1988), 407–418.
[39] L. Puchinger, H. Stachelberger, Microsampling and microscopal assessment of parchment including microchemical analysis of binding media, dyes and inks, in: R. Larsen (Ed.), *Microanalysis of Parchment*, Archetype Publications, London, pp. 3–8, 2002.
[40] T.J. Wess, M. Drakopoulos, A. Snigirev, J. Wouters, O. Paris, P. Fratzl, M. Collins, J. Hiller, K. Nielsen, The use of small angle X-ray diffraction studies for the analysis of structural features in archaeological samples, *Archaeomerty*, **43** (2001), 117–129.
[41] C.J. Kennedy, J.C. Hiller, M. Odlyha, K. Nielsen, M. Drakopoulos, T.J. Wess, Degradation in historical parchments; structural, biochemical and thermal studies, *PapierRestaurierung*, **3** (2002) 4, 23–30.
[42] C.J. Kennedy, M. Vest, M. Cooper, T.J. Wess, Laser cleaning of parchment: structural, thermal and biochemical studies into the effect of wavelength and fluence, *Applied Surface Science*, **227** (2004b), 151–163.
[43] P.L. Privalov, E.I. Tiktopoulo, Thermal conformational transformation of tropocollagen. I. Calorimetric study, *Biopolymers*, **9** (1970), 127–130.
[44] R. Larsen, M. Vest, K. Nielsen, Determination of hydrothermal stability (shrinkage temperature) of historical leather by the micro hot table technique, *Journal of Leather Technologists and Chemists*, **77** (1993), 151–56.
[45] K.A. Piez, Soluble Collagen and Denaturation Products C: Cyanogen Bromide Cleavage, in: G.N. Ramachandran (Ed.), *Treatise on Collagen*, Academic Press, London and New York, pp. 244–246, 1967.
[46] M. Burghammer, M. Muller, C. Riekel, X-ray synchrotron radiation microdiffraction on finrous biopolymers like cellulose and in patricular spider silks, *Recent Research Developments in Macromolecules*, **7** (2003), 103–125.
[47] C. Chahine, Changes in hydrothermal stability of leather and parchment with deterioration: a DSC study, *Thermochimica Acta*, **365** (2000), 101–110.
[48] S.D. Gorham, N.D. Light, A.M. Diamond, M.J. Willins, A.J. Bailey, T.J. Wess, N.J. Leslie, Effect of chemical modifications on the susceptibility of collagen to proteolysis. II. Dehydrothermal crosslinking, *International Journal of Biological Macromolecules*, **14** (1992), 129–138.
[49] R. Larsen, D.V. Poulsen, D.J. Bechmann, M. Vest, B.V. Hansen, K. Nielsen, S. Boghosian, F. Juchauld, H. Jerosch, M. Odlyha, G. Della Gatta, E. Badea, A. Schiraldi, D. Fessas, J. Vnoucek, J. Palm, J.C. Hiller, C.J. Kennedy, T.J. Wess, *History Goes Jelly*, (2006) in preparation".
[50] A. Snigirev, V. Kohn, I. Snigireva, E. Aristova, A compound refractive lens for focusing high-energy X-rays, *Nature*, **384** (1996), 49–51.
[51] T.J. Wess, I. Alberts, J. Hiller, M. Drakopoulos, A.T. Chamberlain, M. Collins, Microfocus small angle X-ray scattering reveals structural features in archaeological bone samples: detection of changes in bone mineral habit and size, *Calcified Tissue International*, **70** (2002), 103–110.
[52] A. Bigi, M. Burghammer, R. Falconi, M.H. Koch, S. Panzavolta, C. Riekel, Twisted plywood pattern of collagen fibrils in teleost scales: an X-ray diffraction investigation, *Journal of Structural Biology*, **136** (2001), 137–143.
[53] K.M. Meek, C. Boote, (2004) The organization of collagen in the corneal stroma. *Experimental Eye Research*, 78, 503–512.
[54] A.-G. Ascenzi, A. Ascenzi, B. Benvenuti, M. Burghammer, S. Panzavolta, A. Bigi, Structural differences between "dark" and "bright" isolated human osteonic lamellae, *Journal of Structural Biology*, **141** (2003), 22–33.
[55] C.J. Kennedy, J.C. Hiller, D. Lammie, M. Drakopoulos, M. Vest, M. Cooper, W.P. Adderley, T.J. Wess, Microfocus X-ray diffraction of historical parchment reveals variations in structural features through parchment cross sections, *Nano Letters*, **4** (2004a), 1373–1380.
[56] A. Somogyi, M. Drakopoulos, L. Vincze, B. Vekemans, C. Camerani, K. Janssens, A. Snigirev, F. Adams, ID18F: a new micro X-ray fluorescence end-station at the European Synchrotron Radiation Facility (ESRF): preliminary results, *X-ray Spectrometry*, **30** (2001), 242–252.

[57] B. Vekemans, K. Janssens, L. Vincze, F. Adams, P. Van Espen, Comparison of several background compensation methods useful for evaluation of energy-dispersive X-ray fluorescence spectra, *Spectrochimica Acta*, **50B** (1995), 149–169.
[58] J.G. Bloodworth, M.J. Parkinson, The display of parchment and vellum, *Journal of the Society of Archivists* **9** (1988), 65–68.
[59] M. Cooper, *Laser Cleaning in Conservation*, Butterworth Heinemann, Oxford, UK, 1997.
[60] W. Kautek, S. Pentzien, A. Conradi, D. Leichtfried, L. Puchinger, Diagnostics of parchment laser cleaning in the near-ultraviolet and near-infrared wavelength range: a systematic scanning electron microscopy study, *Journal of Cultural Heritage*, **4** (2003), 179–184.
[61] M. Cooper, S. Sportun, A. Stewart, M. Vest, R. Larsen, D. Poulsen, Laser cleaning of an eighteenth-century parchment document, *The Conservator*, **24** (2000), 71–78.
[62] W. Kautek, S. Pentzien, M. Rollig, P. Rudolph, J. Kruger, C. Maywald-Pitellos, H. Bansa, H. Grosswang, E. Konig, Near-UV laser interaction with contaminants and pigments on parchment: laser cleaning disgnostics by SE-microscopy, VIS- and IR-spectroscopy, *Journal of Cultural Heritage*, **1** (2000), S233–S240.
[63] S. Sportun, M. Cooper, A. Stewart, M. Vest, R. Larsen, D.V. Poulsen, An investgation into the effect of wavelength in the laser cleaning of parchment, *Journal of Cultural Heritage*, **1** (2000), S225–S232
[64] W. Kautek, S. Pentzien, P. Rudolph, J. Kruger, E. Konig, Laser interaction with coated collagen and cellulose fibre composites: fundamentals of laser cleaning of ancient parchment manuscripts and paper, *Applied Surface Science*, **127–129** (1998), 746–754.
[65] J.G. Grossmann, Shape Determination of Biomolecules in Solution from Synchrotron X-ray Scattering, in: P. Sabatier, E.R. Pike (Eds.), *Scattering*, Academic Press, pp. 1123–1139, 2002.
[66] A. Guinier, *X-ray diffraction in crystals, imperfect crystals and amorphous bodies*, Freeman and Sons, San Francisco and London, 1963.
[67] A. Guinier, G. Fournet, *Small-Angle Scattering of X-Rays*, Wiley, New York, 1955.
[68] Y.W. Zheng, B.S. Luk'yanchuk, Y.F. Lu, W.D. Song, Z.H. Mai, Dry laser cleaning of particles from solid substrates: experiments and theory, *Journal of Applied Physics*, **90** (2000), 2135–2160.

## Chapter 5

# Egyptian Eye Cosmetics ("Kohls"): Past and Present

A.D. Hardy[1], R.I. Walton[2], R. Vaishnav[3], K.A. Myers[4], M.R. Power[5] and D. Pirrie[5]

[1]*Centre for Medical History, School of Humanities and Social Sciences, University of Exeter, Exeter EX4 4RJ, Devon, UK*
*Email: a.d.hardy@exeter.ac.uk*
[2]*Department of Chemistry, the Open University, Walton Hall, Milton Keynes, MK7 6AA, UK*
[3]*College of Medicine, Sultan Qaboos University, Box 35, Al-Khod 123, Oman*
[4]*Department of Chemistry, School of Biological and Chemical Sciences, (now: School of BioSciences) University of Exeter, Exeter EX4 4QD, Devon, UK*
[5]*Camborne School of Mines, University of Exeter in Cornwall, Tremough Campus, Penryn TR10 9EZ, Cornwall, UK*

**Abstract**

The published literature was summarised and reviewed for historical/archaeological data on the usage and chemical composition of Egyptian eye cosmetics ("kohls"). A total of 27 kohl samples were purchased in modern-day Egypt; 18 in Cairo, 4 in Aswan and 5 in Luxor. Also, very small amounts of material were carefully removed from inside six Pharaonic kohl pots (held in the Royal Albert Memorial Museum and Art Gallery, Exeter, UK). These pots were dated to either Middle or New Kingdom (*i.e.* between c. 2040 BC to c. 1070 BC). Each of the 33 samples was analysed by one or more of the following techniques: X-ray powder diffraction (XRPD), low vacuum scanning electron microscopy (LV SEM), infrared spectroscopy (IR) and the relatively new technique of quantitative scanning electron microscopy (QEMSCAN). For the 27 modern-day samples, it was found that the main component for six was galena (PbS) and for the remaining samples one of the following: amorphous carbon (6), calcite ($CaCO_3$) (6), elemental silicon (1), talc ($Mg_3Si_4O_{10}(OH)_2$) (2), cuprite ($Cu_2O$) (1), goethite (FeO(OH) (1), barite ($BaSO_4$) (1), halite (NaCl) (1), and for two samples, an unknown amorphous organic compound. Three of the galena-based samples also had their (average) particle sizes determined from electron micrographs. Five of the six Pharaonic kohl pots contained lead-based compounds (one being empty of kohl). Two were black and so were most likely galena, whilst the other three were white and so were one or more of several possible "made" or natural lead compounds. The composition of the six Pharaonic kohl pots was also studied (only using the LV SEM and QEMSCAN analytical techniques) and the results compared to the compositions expected from previous visual inspections. Whilst only about a quarter (22%) of the modern-day Egyptian kohl samples contained a lead compound as a major component; it was seen, both from our results and those reviewed, that in the Pharaonic past this percentage was much higher.

**Keywords:** Pharaonic Egypt, eye cosmetics, kohl, X-ray powder diffraction, low vacuum scanning electron microscopy, quantitative scanning electron microscopy.

**Contents**



Physical Techniques in the Study of Art, Archaeology and Cultural Heritage
Edited by D. Bradley and D. Creagh

## 1. INTRODUCTION

In ancient Egypt, even sacrificial cows had eye make-up applied before their ritual slaughter. A relief in the fifth dynasty (*i.e.* c. 2400 BC) temple of King Sahure shows such cows being tended to by female temple personnel [1]. As for the details of the usage of eye make-up/ eye-paint ("kohl") by the human population of ancient Egypt, there is sometimes a degree of uncertainty with respect to the dates of usage, exact original composition of the kohls and their reason(s) for usage (*i.e.* if cosmetic and/or medicinal and/or magico-religious and if used as cosmetics, were they used on the eye and/or face). However, all the authors agree that there was green eye-paint initially and "a little later" black eye-paint and that it was used for one or more of the above reasons by men, women and children from all social levels.

The green eye-paint was usually the ore malachite, basic copper carbonate, which was mined in Sinai and the Eastern desert. It has been found in Badarian/pre-dynastic (*i.e.* 5000–3000 BC) tombs (Ref. [2]; an unusual example, involving the burial of an elephant) and until recently was assumed to have been used until at least the nineteenth dynasty (*i.e.* c. 1250 BC) [3,4]. However, a more recent publication [5] states that the green (malachite) eye-paint "... seems to have been used only until the middle of the old kingdom, when it was replaced by the black galena-based form ...". The middle of the old kingdom was c. 2500 BC.

Black eye-paint has been found at least once in a Badarian period tomb, occurs more often in later tombs and continued to be in use until the Coptic period (*i.e.* c. 395–640 AD). The term "msdmt"/"mesdemet" was used for eye-paint in general and for black eye-paint in particular; whereas the green eye-paint was referred to by the term "wadju"/"udju"/ "ouadjou". Results of many analyses of tomb funerary gifts of black/grey-black eye-paint show that often the main component was the ore galena (lead sulphide). This ore was mined in several localities in Upper Egypt. Sometimes, malachite and galena were also imported, often from Arabia *via* "Punt" (which was probably the present-day Eritrea/Somalia). Both eye-paints occur, either individually or together, in several of the listed (papyri) therapeutic "recipes" used as eye salves by the ancient Egyptians [6]. For those readers interested in the details of ancient Egyptian medicine, which is beyond the scope of this chapter, there are several books/reviews/articles etc. available in the published literature [7–9]. Also, for an overview of eye diseases in ancient Egypt, see Ref. [10].

> "Cleopatra experimented with eye colours to effect, using black galena on the eyelid and to delineate her brows, and painting the upper lids deep blue, and the lower lids bright green" [11]

There are "statements", such as the one above, in the published literature of Cleopatra's usage of a *blue* eye cosmetic. Sometimes it is merely stated that the blue eye cosmetic was used by (Pharaonic) "Egyptian ladies" and sometimes, additionally, that it was lapis (lazuli) [12]. No source reference(s) for any of these statements have ever been seen by these authors.

Cleopatra VII (69–30 BC) may well have used blue eye cosmetic to entrance Mark Antony, but stating that she used lapis lazuli for this effect is pure speculation. Lapis lazuli *was* used for small objects (*e.g.* beads, amulets, scarabs, etc.) from predynastic times in Egypt and most probably it came from the remote mines of Badakshan (north-east Afghanistan). It was also used for inlay jewellery from predynastic times. The inlaid eyelids and eyebrows of Tutankhamun's famous gold mask are made of lapis lazuli. However, such usage should not be taken to mean that the material was used as an ingredient for eye cosmetics. It was also used as a paint pigment (in ultramarine), but much later (c. eleventh century AD) [3]. It is also occasionally listed in (papyrus) recipes for eye diseases/infections [6].

We have come across only a few publications in English which give data on the numbers, provenance (*i.e.* data on the date/period and name of the site of excavation) and compositions from chemical analytical studies, of the different eye-paint/makeup/cosmetics found in ancient Egyptian tombs. One such reference (Ref. [3]) summarises 74 such samples, from analytical work done by both himself (Lucas) and several prior authors. Some analytical results (*i.e.* those of Ref. [13]) were deliberately excluded by these authors for a variety of reasons (such as: missing dates and origins, and missing numbers of the types of particular samples studied). Also, whilst 73 of the 74 samples summarised were free from resin, some of the excluded samples did contain such material and so were regarded by these authors as not being cosmetic samples. Wax and/or fatty matter was found in a few of the 74 summarised samples and these were regarded as probably being cosmetics of some sort.

The above 73 results are given in Table 1. We have excluded, from the original 74 summarised samples, the one sample whose analysis was listed as "uncertain". It can be seen from this table that the majority of the grey/grey-black/ black samples are galena-based; that is 44 of 58, *i.e.* 76%. Also, the only antimony (tri) sulphide-based (*i.e.* the ore stibnite as the main component) sample found was dated to the nineteenth dynasty, whilst the 44 galena-based samples had a wide variety of dates. Thus Table 1 represents a summary of the analyses on cosmetics (assumed by us to be mostly kohls/eye-paints) samples carried out (by a variety of authors) from the 1880s to the early 1960s; where the analyses were mostly performed using "wet chemistry" analytical techniques.

In contrast, using modern analytical (*i.e.* synchrotron, spectroscopic) equipment, it has recently been shown by Louvre-based researchers [14–17] that several *synthesised* lead compounds (phosgenite ($Pb_2Cl_2CO_3$) and laurionite (PbOHCl)) were deliberately made and then added to usually galena-based cosmetics for either their (supposed) therapeutic

**Table 1.** Composition of eye-paints ("kohls") from ancient Egypt (1)[a]

| Number | Main component | Additional components |
|---|---|---|
| 44 | Galena (PbS) | 1 with Gypsum ($CaSO_4.2H_2O$); 5 with Carbon; 2 with trace of Antimony Sulphide ($Sb_2S_3$) |
| 2 | Carbonate of lead ($PbCO_3$) | 1 with trace of $Sb_2S_3$ |
| 6 | Brown Ochre ($Fe_2O_3.nH_2O$) | None |
| 1 | Limonite (Iron oxide/hydroxide mixture) | None |
| 2 | Magnetic oxide of iron ($Fe_3O_4$) | 1 mixed with earthy matter |
| 10 | Oxide of manganese ($MnO_2$) | None |
| 1 | Black oxide of copper (CuO) | None |
| 1 | Sulphide of antimony ($Sb_2S_3$) | None |
| 5 | Malachite ($Cu(OH)_2.CuCO_3$) | 1 mixed with resin |
| 1 | Chrysocolla ($CuSiO_3.2H_2O^b$) | None |

[a] From Ref. [3].

[b] One formula for this mineral, others exist.

effects or possibly to give varying shades of colour to them for use as eye/face cosmetics. The latter effect was also achieved by adding a (white) available lead ore – cerussite (lead carbonate). It is thought that these light grey or white mixtures may have been used as face cosmetics or even as foundation creams. Some of their analytical results are summarised in Table 2. In this table we have included only those published results where both complete (*i.e.* the percentages quoted do sum to 100%) compositional and time period data have been given for the samples studied. Many of these results show that "fatty acids of animal provenance" *are* also present and hence there is some discussion as to their exact usage by the ancient Egyptians (see below). These "made additions" appear to have been initiated at around 2000 BC and to have continued to at least c. 1200 BC.

As for any analyses of ancient Egyptian kohls from the early 1960s (*i.e.* after the analyses summarised by Lucas and Harris [3]) to the present (but apart from the above mentioned Louvre-based work) – we have found no publications describing any such analyses. However, there have been several papers published where kohls were one of several lead-based ancient Egyptian artefacts subjected to lead isotopic analysis (LIA). All the kohls (29 samples) so studied contained lead to varying degrees and all are stated to contain galena. Seventeen kohls (1 lump and 16 powders, variously dated from the sixth to eighteenth dynasties) were studied by Brill *et al.* [18]; eleven (mostly powders, and all pre-/protodynasty dated) by Stos-Gale and Gale [19] and one ("clots of galena" found at their excavation site and radiocarbon dated to 3080 ± 110 BC) by Hassan and Hassan [20]. No details of any chemical analyses undertaken on the kohl samples are given in any of these three publications. As almost all the above samples were obtained from UK museums,

**Table 2.** Composition of eye-makeup ("kohls") from ancient Egypt (2)[a]

| Sample number | Pharaonic period | Major component (%) | Minor components (%) | Reference |
|---|---|---|---|---|
| N811d | New Kingdom | Galena (PbS) (75) | Cerussite ($PbCO_3$) (14); Phosgenite ($Pb_2Cl_2CO_3$) (11) | [14] |
| N811g | New Kingdom | Galena (75) | Cerussite (14); Phosgenite (11) | [14] |
| N1332 | New Kingdom | Galena (89) | Phosgenite (6); Cerussite (3); Laurionite (PbOHCl) (2) | [15] |
| N1367d | New Kingdom | Galena (100) | None | [15] |
| N1367g | New Kingdom | Galena (100) | None | [15] |
| AF167 | New Kingdom | Galena (100) | None | [16][b] |
| AF6772 | New Kingdom | Galena (62) | Cerussite (28); Laurionite (10) | [16][b] |
| E11047 | New Kingdom | Galena (41) | Cerussite (41); Phosgenite (18) | [14] |
| E11048b | New Kingdom | Galena (50) | Phosgenite (37); Cerussite (13) | [16][b] |
| E11048c | New Kingdom | Cerussite (42) | Phosgenite (29); Galena (20); Anglesite ($PbSO_4$) (9) | [15] |
| E11048d | New Kingdom | Cerussite (47) | Galena (28); Phosgenite (25) | [15] |
| E11048e | New Kingdom | Galena (40) | Phosgenite (28); Cerussite (26); Laurionite (3); Anglesite (3) | [15] |
| E14455 | New Kingdom | Galena (70) | Zinc-based cpds. (20); Anglesite (10) | [14] |

*Continued*

**Table 2.** Continued

| Sample number | Pharaonic period | Major component (%) | Minor components (%) | Reference |
|---|---|---|---|---|
| E14569 | New Kingdom | Laurionite (35) | Cerussite (25); Galena (24); Phosgenite (16) | [16][b] |
| E20514 | New Kingdom | Galena (72) | Phosgenite (9); Sphalerite (ZnS) (9); Anglesite (4); Cerussite (3); Smithsonite ($ZnCO_3$) (2); Laurionite (1) | [17][b]; [15] |
| E21562 | Middle or New Kingdom | Phosgenite (32) | Cerussite (25); Laurionite (19); Galena (15); Anglesite (9); | [14] |
| N1209 | Middle Kingdom | Galena (44) | Anglesite (21); Phosgenite (10); Cerussite (13); Laurionite (12) | [15] |
| E23100 | Middle Kingdom | Galena (58) | Anglesite (19); Cerussite (19); Laurionite (4) | [14] |
| E23105 | Middle Kingdom | Galena (44) | Quartz ($SiO_2$) (38); Calcite ($CaCO_3$) (18) | [14] |

[a] Based on various published papers, all of which were supported by L'Oreal Research and the Louvre Museum, France.

[b] The references where the finding of "fatty acids" in the samples is mentioned.

it is possible that some have been previously analysed and are amongst those summarised by Lucas and Harris [3].

Both ancient Egyptian eye-paints, green and black, were initially ground to a powder and then applied dry or were mixed with water or a water-soluble gum to give a paste which was then applied with a finger or, at later dates, by means of a kohl-stick. This stick was made of stone (*e.g.* black haematite), bone, wood or ivory. It is still a matter of some discussion if fatty-matter/oil/wax, or even possibly resin, was separately used for applying the kohl or if the presence of this additional material, when actually within the ancient funerary containers, meant that it was another form of cosmetic (*i.e.* for the face, the fats etc. giving good coverage/spreadability) or if it was to be used as an eye salve in the after-life.

In the "Sayings of the Prophet" (PBUH) Muslims are advised to use kohl made from "ithmid" (or sometimes, "ethmid"/"athmid"). In Hadith number 1, it is stated " … 'Use kuhl made from ithmid on the eye; it brightens eyesight and strengthens the growth of the eye lashes' … " (www.ummah.net/moa-on-line/hadith/shumaail/st7.html). "Ithmid" was described as being a dark, reddish-black, shiny stone and has always been assumed to be the ore stibnite. Possibly, over time, the ore galena was substituted; as galena was not as uncommon, looked and felt much the same and was much less expensive to obtain (also see in Section 4.1).

Some of the early Muslim medical authors made a clear distinction between two types of eye powders/pastes: "siyaf" powders and pastes which were for cosmetic purposes and which were usually made from soot (*i.e.* amorphous carbon), whilst "kuhl" (kohl) powders and pastes were used to treat various conditions and diseases of the eye. The latter were, and still are, made in southern Oman from the charcoal obtained from the root of the local "ra" (*Aerva javanica*) plant, plus small amounts of mother-of-pearl and frankincense and, most importantly, some antimony/lead sulphide (*i.e.* the "ithmid"; this being perceived as being the "active ingredient") [21].

This distinction, of two types of eye powders/pastes, was still being observed in Egypt of the 1820s and 1830s. It was reported that cosmetic kohl was obtained from burning an aromatic resin ("liban", a species of frankincense, see later) or shells of almonds; but that kohl with medicinal properties was made by mixing galena powder with Sarcocolla (mother-of-pearl), long pepper (a particularly potent variety of pepper), sugar candy, fine dust of Venetian sequin (an old gold coin), and sometimes powdered pearl [20,22].

In modern-day Egypt, kohls are used only by females, and their children, from all socio-economic classes, for both beautification and as an ethnic remedy, *i.e.* to relieve eyestrain, pain or soreness. Anecdotal evidence (Hardy; personal communications, 2001) suggests that there is more use of kohl in rural than urban areas of Egypt, and that middle/upper class ladies in urban areas such as Cairo are now-a-days more likely to use the modern eye pencils for reasons of convenience of purchase, less messy in usage and because they are "known to be good". Also, "traditional" kohls are perceived as "old fashioned", especially by the younger student generation.

One "traditional" recipe for making home-made kohl in modern-day Cairo is (Hardy; personal communications, 2001) (extra data has sometimes been added, in *italics*): " … add olive oil to "leban dakar" (*this is the Egyptian vernacular name for gum-resin in general and for frankincense in particular; see above for its usage in Egypt of the 1800s*) until it is covered (in a saucer). Leave for 2 days and then transfer the contents to a plate. Add several small pieces of cotton and then light them; as it burns, place a pottery bowl over it to collect the smoke residue (*soot*). Leave for an hour. Remove the smoke residue from the bowl with a feather and place in a kohl jar". It is then applied daily using an applicator stick, which is usually made of glass or plastic. Sometimes olive oil is used for ease of actual application of the kohl. This "recipe" can be compared to another reported method of preparing home-made kohl in modern-day Egypt [23]: "... burn the following ingredients together – sumach (*a shrub with green flowers and red hairy fruits*), nutmeg tree, extracts from previously burnt frankincense, sugar crystals and perfumed cherry". The resulting ash is then used as the kohl. Both of these "recipes" give essentially the same end product – innocuous amorphous carbon.

## 2. MATERIALS AND METHODS

### 2.1. Modern-day samples

A total of 18 samples were obtained from the city markets (souks) of Cairo. The cost of these kohls varied between 2.5 and 10 Egyptian pounds, where one GBP was then 5.5 Egyptian pounds. Later, nine additional samples of mostly *coloured* kohls were purchased in the souks of Aswan (4 samples) and Luxor (5 samples). Their price varied between 1 and 10 Egyptian pounds. Because of the brief duration of these later visits to Aswan and Luxor, it was decided beforehand to look mostly for and purchase "unusual" (*i.e.* coloured) kohl samples. All the (few) available green, blue or red samples seen were purchased. Figure 1 shows how these 27 kohls are distributed by the countries where they were made.

All these kohl samples were examined by the analytical techniques of X-ray powder diffraction (XRPD) and scanning electron microscopy (SEM) with an attached energy dispersive X-ray microanalyser. Additionally, ten of the (artificially) coloured samples were examined by infrared (IR) spectroscopy.

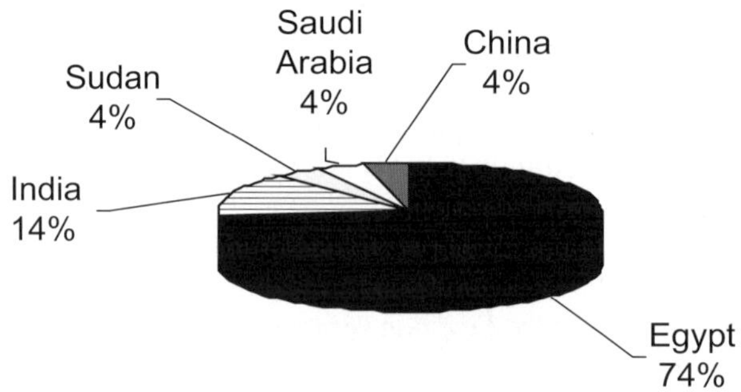


**Fig. 1.** Orgins of modern-day Egyptian kohl samples studied.

### 2.1.1. *X-ray diffraction*

The kohl samples were, where necessary, ground to a powder and then mounted in an aluminium holder. The Cairo samples were analysed using a Bruker AXS D8 Advance diffractometer. Diffraction data was collected with generator settings of 40 kV and 40 mA. A monochromator and automatic divergence slits were used with the CuK$\alpha$ X-ray radiation. One data scan was performed for each sample, using a step scan of size 0.02° and a time of 2 s/step over a 2-$\theta$ range of 2–80°. The Aswan and Luxor (powder) samples were analysed using a Siemens D500 diffractometer, using CuK$\alpha$ X-ray radiation. One data scan was performed for each sample, using a step scan size of 0.0256° and a time of 1 s/step over a 2-$\theta$ range of 8–70°.

These scans were used to determine the major and minor components (phases) present in the samples by comparing the obtained data to the reference data in the 2000 JCPDS (Joint Committee for Powder Diffraction Standards) database [24]. The major phase was defined to be that phase which had a presence of >80% of the sample. Semi-quantitative results were obtained using this analytical method; any percentages quoted are estimates only. The minor phases given in brackets in Tables 3 and 4 are those estimated to be ≤5% of the sample. Also, each group of minor phases are listed in decreasing order of their estimated percentage presence.

### 2.1.2. *X-ray microanalyser*

Each of the kohl (powder) samples was mounted on an aluminium stub using an adhesive carbon tab. They were then examined in a JEOL JSM 5300 LV SEM with a Rontec energy dispersive X-ray (EDX) microanalyser attached.

The LV (low vacuum) SEM is designed so that its electron gun and electron optical system are kept under high vacuum (typically about a millionth of a torr), whilst the specimen chamber is differentially evacuated to low vacuum (typically a few torr) by another pumping system. The gas molecules surrounding the electron beam are ionised and the electric charge on the specimen is neutralised, thus allowing non-conductive specimens to be studied without coating.

Elements lighter than carbon (*i.e.* $Z \leq 5$) cannot be detected using the above equipment. Detection was qualitative and the element peaks that were only just above the background are given in brackets in Tables 3 and 4. Three Cairo samples (nos. 1, 3 and 6 in Table 3) were imaged at various magnifications in back-scattered mode, so that an estimate could be made of the average particle size of the galena cubes in each sample.

### 2.1.3. *Infrared spectroscopy*

Infrared spectra were collected on ten of the kohl samples which were artificially coloured, *i.e.* those coloured with a small amount of a (probably synthetic) colourant and not from the colour of the major phase (or even one of the listed minor phases). These were the 7 (of the 9, that is all of them except L4 and L5) samples from Aswan and Luxor (see Table 4) and three samples from Cairo (nos. 12, 13 and 14 in Table 3). The spectra were collected using a Perkin Elmer IR spectrometer, with each sample made into pellet form after mixing with potassium bromide.

## 2.2. Pharaonic samples

Recently six Pharaonic kohl pots became available for study by us. Small amounts ("scrapings") were carefully removed from all six pots, both from the inside (for the kohl; sample nos. M1 to M6) and the outside (for pot material; sample nos. MP1 to MP6). Table 6 summarises the data (*e.g.* Museum catalogue numbers, approximate age, possible provenance, height, texture, colour and initial analytical results) on the pots and their contents. Figure 4 shows all six pots with a centimetre scale (with a colour picture of them on the book's front cover).

All the pots belong to the Royal Albert Memorial Museum (Exeter, UK) and the Museum's knowledge of their provenance etc. is unfortunately limited by the paucity of the data provided by the original donors. However, we have been able to add a little more information (Morkot; personal communication, 2005; see Table 6). Namely that pots MP1, MP2, MP4 and MP5 are probably of Middle Kingdom date, whilst MP3 is probably of New Kingdom date. Pot MP6 can be dated with near certainty to the Middle Kingdom because of its obvious "Blue marble" composition. Also, the compositions of MP4 and MP5 can also be stated, with near certainty, to be "Egyptian alabaster". However, the listed compositions for MP1, MP2 (both as travertine) and for MP3 (as "Egyptian ceramic") are tentative. Pots MP1, MP2 and MP3 are part of the Montague collection in the Museum, whilst pots MP4, MP5 and MP6 are supposed to have been loaned to them from the Petrie Museum of Egyptian Archaeology (London, UK) some decades ago.

The samples were examined by the analytical techniques of SEM with an attached energy dispersive X-ray microanalyser and by automated scanning electron microscopy with linked energy dispersive spectrometers (QEMSCAN).

### *2.2.1. X-ray microanalyser*

Nine of the twelve Pharaonic samples were mounted on individual aluminium stubs using adhesive carbon tabs. They were then examined in a JEOL JSM 5300 LV SEM with a Rontec energy dispersive X-ray (EDX) microanalyser attached, as for the modern-day samples above.

Element detection was both qualitative and quantitative and the elements found are given, in decreasing order of their weight percent, in Table 6. The elements that were at less than 1% each are given in brackets. As the amounts of samples MP1, MP2 and MP3 were extremely limited, it was decided to do QEMSCAN analyses only on them (see below).

### *2.2.2. QEMSCAN (quantitative scanning electron microscopy)*

QEMSCAN is an automated scanning electron microscope, which provides particle-by-particle quantitative mineral data [25–27].

The QEMSCAN operating system comprises a scanning electron microscope coupled with four energy dispersive spectrometers arranged at approximately 90° intervals around the sample chamber.

The particle mineralogical analysis (PMA) mode of operation was used for the analyses of these 12 samples. This mode of analysis is conducted where the particles are automatically located, using the contrast in backscatter coefficient (which is proportional to the mean atomic mass of the material) between the particle and the mounting substrate (which was epoxy resin for all these samples). Once located, the electron beam is rastered across the particle at a user-defined stepping interval (pixel spacing). Here, for the pot contents (*i.e.* the kohls; sample nos. M1 to M6), two particle size fractions were analysed. Particles between 1 and 25 μm in size were analysed using a pixel spacing of 0.5 μm and particles between 25 and 200 μm in size were analysed using a pixel spacing of 2 μm (see Table 7). For the pot materials themselves (*i.e.* samples nos. MP1 to MP6), analysis was done on particles of between 25 and 200 μm in size at a pixel spacing of 2 μm (see Table 8).

At each pixel spacing, an X-ray energy spectrum was rapidly acquired and compared with a look-up table of known chemical composition minerals (the Species Identification Protocol (SIP)), then a mineral identification is made, and its weight percent is subsequently calculated. Individual identifications are typically made online in under 1 ms, which equates to in excess of 250 000 individual analyses points per hour (taking into account the time needed to move the sample stage and electron beam). All analytical measurements are stored, thus allowing the sample to be re-interpreted subsequently off-line.

As only small/very small amounts of material were available for QEMSCAN analysis, the individual samples were prepared as "sprinkle particle mounts". This involved collecting a tiny amount of sample on the end of a clean cocktail stick and dispersing the particles in a drop of methanol on a polished blank epoxy resin block. The methanol was allowed to flash off and the sample was coated with carbon to a thickness of 250 Å.

## 3. RESULTS

### 3.1. Modern-day samples

Table 3 lists the results obtained, for the 18 modern-day Cairo samples, in the order: (first) lead-based, amorphous carbon/carbon-based (in an amorphous organic compound), calcium-based, copper-based, iron-based and (last) silicon-based (for the main element of the major phase present) samples.

Six of these eighteen samples studied were lead-based; the four purchased as powders were black or grey-black in colour whilst the two lumps were both silver-grey in colour. The major phase was always found to be galena (PbS) and all the powders contained minor phases of anglesite ($PbSO_4$) and cerussite ($PbCO_3$). All except one of these six samples have an estimated percentage for galena of ≥95%. The exception, sample no. 2, contains about 10% of camphor ($C_{10}H_{16}O$) and about 3% of zincite (ZnO) in addition to the above mentioned minor phases of anglesite and cerussite (each about 1%). It thus contains only about 85% of galena. Of these six lead-based samples, two originated in India and four in Egypt. Two of the latter four were purchased as lumps ("kohl hagar", that is "kohl stone") and both contained very small amounts (*i.e.* ≤1%) of anglesite. One of these two (sample no. 3), obtained in a well-established shop in the main souk (Khan Al-Khalili) of Cairo, was insisted to be (by the shop-keeper), *wrongly*, the ore stibnite (that is antimony trisulphide, $Sb_2S_3$). This again highlights the very similar appearance (metallic lustre), colour (silver-grey/grey-black) and feel (relatively soft) of the ores galena and stibnite, especially when they are in massive states.

**Table 3.** Analysis of kohl samples from modern-day Cairo

| Sample no. | Texture | Colour | Made in | Purchased in | XRPD major phase | XRPD minor phase(s)[d] | SEM[c] |
|---|---|---|---|---|---|---|---|
| 1 | Powder | Black | India (Bombay) | Cairo main souk[a] | Galena (PbS) | (Anglesite ($PbSO_4$)) (Cerussite ($PbCO_3$)) | Pb, S, C, O |
| 2 | Powder | Grey-black | India (Bombay) | Cairo main souk | Galena | Camphor ($C_{10}H_{16}O$) (Zincite (ZnO)) (Anglesite) (Cerussite) | Pb, S, C, O, Zn |
| 3 | Lump | Silver-grey | Egypt | Cairo main souk | Galena | (Anglesite) | Pb, S, C, O |
| 4 | Lump | Silver-grey | Egypt | A Cairo souk (Shubra) | Galena | (Anglesite) | Pb, S, C, O |
| 5 | Powder | Grey-black | Egypt | Cairo main souk | Galena | (Anglesite) (Cerussite) | Pb, S, C, O |
| 6 | Powder | Grey-black | Egypt | Cairo main souk | Galena | (Anglesite) (Cerussite) | Pb, S, C, O |
| 7 | Powder | Black | Egypt | Cairo main souk | Amorphous carbon | None | C, O, S |
| 8 | Powder | Black | China | A Cairo souk (Shubra) | Amorphous carbon | None | C, O, N (S) |
| 9 | Powder | Black | India (Bombay) | Cairo main souk | Amorphous carbon | (Talc ($Mg_3Si_4O_{10}(OH)_2$)) (Quartz ($SiO_2$)) | C, Si, Mg, O, Cl (S) |

| | | | | | | | |
|---|---|---|---|---|---|---|---|
| 10 | Powder | Black | Egypt | Cairo main souk | Amorphous carbon | None | C, S, O |
| 11 | Powder | Black | India | A Cairo souk (Shubra) | Amorphous carbon | (Zincite) | C, S, O, Zn |
| 12 | Powder | Blue | Egypt | Cairo main souk | An unidentified amorphous cpd | Unknown | C, Cl, S, O (N, Si) |
| 13 | Powder | Green | Egypt | Cairo main souk | An unidentified amorphous cpd | Unknown | C, S, O, N |
| 14 | Powder | Purple | Egypt | Cairo main souk | Calcite ($CaCO_3$) | Talc and unidentified | Ca, O, Si, C, S, Na, Mg, Ti, Fe (Zn, Cu, N) |
| 15 | Lump | Grey-black[b] | Sudan | A Cairo souk (Shubra) | Cuprite ($Cu_2O$) | None | Cu, O, C |
| 16 | Powder | Yellow-brown | Egypt | Cairo main souk | Goethite (FeO(OH)) | Unidentified | Fe, O, C, S, Cu (Si) |
| 17 | Lump | Light grey | Saudi Arabia | Cairo main souk | Silicon | Iron di-silicide ($FeSi_2$) | Si, Fe (C) |
| 18 | Powder | Grey-white | Egypt | Cairo main souk | Talc | Unidentified | Si, Cu, Mg, O, C, Zn |

[a] Khan Al-Khalili.

[b] When ground, a red powder was obtained.

[c] Indicates that the peaks in brackets are only just above background.

[d] Each of the minor phases given in brackets were estimated to be less than 5% level in the sample.

Three of the above lead sulphide-based samples (nos. 1, 3 and 6), had their particle sizes (*i.e.* length of the edge of the cube) visually estimated from SEM images of the galena particles (cubes). Sample no. 1 (a matt black powder) had a size range of only 3–10 μm, and an estimated average size of 5 μm. Sample no. 3 (a silver-grey lump, hand ground to a highly iridescent grey-black powder) had a larger size range, 22–135 μm, and an estimated average size of 69 μm. Sample no. 6 (a slightly iridescent grey-black powder) had a size range of 6–24 μm, and an estimated average size of 11 μm.

Seven of the samples were based on amorphous carbon (5) or on carbon in a coloured amorphous organic compound (2): five were black; one, bright green and one bright blue in colour. Of the other five (black) samples, only two had minor phases; zincite in sample no. 11 and talc ($Mg_3Si_4O_{10}(OH)_2$) and quartz ($SiO_2$) in sample no. 9; both samples originating in India. One of the other black samples originated in China and the remaining two in Egypt. The two brightly coloured samples (nos. 12 and 13) mentioned above were also made in Egypt, had the name "Nefertiti" on their containers and were readily available in the tourist areas of the main souk of Cairo.

One sample (no. 14), also labelled "Nefertiti", was based on calcite ($CaCO_3$) and had a known minor phase of talc (at about 10%). As its colour was purple, it was assumed to have a small amount of an unknown colourant present. The sample (no. 16) based on goethite (FeO(OH)) and the one (sample no.18) based on talc were yellow-brown and grey-white in colour respectively. Both contain small amounts of at least one minor phase each, the nature of which are currently unknown. Their colours correspond to the natural colours of the two major phases found. Their containers, the prices and the shop packaging were all almost identical to the previously mentioned samples that were labelled "Nefertiti". These five samples (nos. 12, 13, 14, 16 and 18), which were all made in Egypt, were regarded by some of the main Cairo souk shopkeepers as being the semi-official kohls and so were the ones most usually offered to tourists.

The IR spectrum of sample 14, purple in colour, showed the peaks for calcite and talc (major and minor phases respectively) and unfortunately these peaks "blotted out" any peaks arising from the small amount of colourant present. However, for samples 12 and 13 (blue and green, respectively, and no phases were identified from the XRPD, as both were totally amorphous), there were good fits for a benzo-sulphonamide compound, possibly with a nitro-group situated "para-" to the sulphonamide group on the benzene ring. These results are supported by the SEM results for these two samples (see Table 3). It is currently unclear if the above sulphonamide compound(s) give the blue and green colours seen for these two samples.

The last two Cairo samples (nos. 15 and 17) were both purchased as lumps, were grey-black and light grey in colour and originated in the Sudan and in Saudi Arabia, respectively. Sample no. 15 gave a red powder on grinding and was found to be pure cuprite ($Cu_2O$) on analysis. This substance has been observed before once before by us, in the context of ethnic remedies/cosmetics of the Middle East; under the name "seika" it was bought in Dubai main souk, where it was sold as a *face* cosmetic [28]. Regarding the other sample (no. 17), it proved to be difficult to grind to a powder and when analysed, the major phase was found to be elemental silicon, with a minor phase of iron di-silicide ($FeSi_2$, at approximately 10%). Such a mixture does *not* occur naturally and how such an unusual, and obviously man-made, material came to be available as a "kohl hagar" in a side-alley of the main souk in Cairo is unclear to us.

Table 4 lists the results of the analyses of the nine kohl samples from modern-day Aswan and Luxor. They are listed in the order of purchase and overall five are based on calcite, one on talc, one on barite ($BaSO_4$), one on halite (NaCl) and one on amorphous carbon. All the calcite-based sample contain approximately 2% quartz each and one (A2) additionally contains approximately 12% talc. The talc-based sample (A1) additionally contains some quartz (11%) and calcite (2%). The barite-based sample (A4) also contains some quartz (2%). Also, 8 of the 9 samples contain a small percentage (1–2%) of a (probably organic) colourant whose exact chemical nature is currently unknown. The IR spectra obtained for 7 (*i.e.* all except L4 and L5) samples had peaks for the major phases (calcite, talc and barite) and, as before for sample 14, these peaks effectively overlapped with any colourant peaks.

### 3.2. Pharaonic samples

Table 6 lists the quantitative elemental data obtained from the LV SEM analyses on nine of the twelve samples (three of the pot-only samples and all six of the internal samples). The last column in the table lists the "Probable main component(s)" for both the eye cosmetic (kohl) contents and the pots themselves; these being based on past experience, colour, the LV SEM results and (for the pots) on the experience of a local archaeologist (Dr. R.G. Morkot; personal communication, 2005).

Pot samples MP1, MP2 and MP3 did not have enough material available to do LV SEM and so their expected chemical compositions of travertine (a naturally occurring calcium carbonate deposit) for MP1 and MP2, and of "Egyptian ceramic" (also known as "Nile mud", mostly a mixture of silicates) for MP3 had to await QEMSCAN analyses (see below). However, for the other three pots, LV SEM was done and its quantitative results (specifically, the first three elements found in each case) supports the listed expected main component in all three cases. That is "Egyptian alabaster" (calcite, $CaCO_3$) for MP4 and MP5, and "Blue marble" (anhydrite, $CaSO_4$) for MP6.

The kohl contents of the six pots show the presence of lead, in significant amounts, in five cases. The one exception, sample M3, showed only a very small amount of lead (less than 1%). Thus pot MP3 was assumed to be empty of kohl and the sample analysed to have been of the pot itself. Two of the pot contents were black (M4 and M5) and this colour plus the presence of both lead and sulphur in significant amounts indicates lead sulphide (the ore galena) to be probably present. The other three analysed contents (M1, M2 and M6) were all white/light brown in colour and were all found to have lead, carbon and oxygen present in significant amounts. Additionally, M1 and M6 each contained chlorine. This indicates that M1 and M6 could contain significant amounts of one or more of the following lead compounds: laurionite (Pb(OH)Cl), phosgenite ($Pb_2(CO_3)Cl_2$), cerussite ($PbCO_3$), and hydro-cerussite ($2PbCO_3.Pb(OH)_2$); while sample M2 (which had no chlorine) could contain significant amounts of only the last two lead compounds mentioned.

Table 7 gives all the QEMSCAN results on the pot contents (samples M1 to M6). The major components for M3 were found to be various silicates (mostly calcium–aluminium silicates), calcite and quartz. Only a very small amount of lead compounds (phases) was found (0.2/0.1%). This confirms the above assumption that pot MP3 is empty of kohl. The other five content samples all have large (*i.e.* >75%) amounts of lead phases present.

**Table 4.** Analysis of kohl samples from modern-day Aswan and Luxor

| Sample no. | Texture | Colour | Made in | Purchased | XRPD major phase | XRPD minor phase(s)[a] | SEM[b] |
|---|---|---|---|---|---|---|---|
| A1 | Powder | Light blue | Egypt | Aswan (souk) | Talc ($Mg_3Si_4O_{10}(OH)_2$) | Quartz ($SiO_2$) (Calcite ($CaCO_3$)) (Colourant*) | Si, Mg, O, C, (Fe, Ca) |
| A2 | Powder | Light blue-green | Egypt | Aswan (souk) | Calcite | Talc (Quartz) (Colourant*) | Ca, C, Si, O, Mg |
| A3 | Powder | Blue | Egypt | Aswan (souk) | Calcite | (Quartz) (Colourant*) | Ca, C, O (Na, S) |
| A4 | Powder | Dark green | Egypt | Aswan (souk) | Barite ($BaSO_4$) | (Graphite) (Quartz) (Colourant*) | Cl, Ba, S, C, Cu, O |
| L1 | Powder | Green | Egypt | Luxor (main tourist souk) | Calcite | (Quartz) (Colourant*) | Ca, C, O (Cl) |
| L2 | Powder | Dark blue | Egypt | Luxor (main tourist souk) | Calcite | (Quartz) (Colourant*) | Ca, C, O |
| L3 | Powder | Dark green | Egypt | Luxor (main tourist souk) | Calcite | (Quartz) (Colourant*) | Ca, C, O (Cl) |
| L4 | Powder | Red | Egypt | Luxor (main tourist souk) | Halite (NaCl) | (Colourant*) | Cl, Na, S, O, C, Fe, K |
| L5 | Powder | Black | Egypt | Luxor (main tourist souk) | Amorphous Carbon | None | C, Ca, S, Si, O |

[a] The minor phases given in brackets were those estimated to be at less than 5% level in the sample.

[b] The elements given in brackets are those whose peaks are only just above background.

* One or more (synthetic) organic colourants, whose exact nature is currently unknown.

**Table 5.** Summary of Egyptian modern-day kohl sample names

| Sample no. | Sample name[a] | Data on contents? | Data on medicinal effects? | Contains lead? |
|---|---|---|---|---|
| (A) Cairo samples: | | | | |
| 1 | Khojati Surma Sada | Y | N | Y |
| 2 | Khojati Surma No. 9 | Y | Y | Y |
| 5 | CHOL NORHAN (cold)[b] | N | N | Y |
| 6 | CHOL NORHAN (hot)[b] | N | N | Y |
| 7 | CHOL NORHAN | N | N | N |
| 8 | Kohl Noori[b] | N | N | N |
| 9 | Hind Ka Noor eye liner | N | Y[b] | N |
| 10 | LUX No. 1 | N | N | N |
| 11 | Black Shahrazad | N | N | N |
| 12 | Nefertiti | N | N | N |
| 13 | Nefertiti | N | N | N |
| 14 | Nefertiti | N | N | N |
| (B) Aswan and Luxor samples: | | | | |
| A2 | Kohl Shahrazed | N | N | N |
| A3 | Cileopatra Super | N | N | N |
| A4 | Cleopatra | N | N | N |
| L1 | Kamal Cleopatra[b] | N | N | N |
| L5 | CHOL MOHGA | N | N | N |

[a] The English name on the label of the container or on the leaflet inside the container, unless translated.

[b] Translated from Arabic.

Unfortunately, at the present time, this analytical technique cannot distinguish between the various lead phases thought to be present (see above). Further work is in progress, using single-element wavelength dispersive spectrometry, and we hope to publish a later paper on the results. Other compounds sometimes found to be present in these five samples, at >2%, were: calcite, gypsum/anhydrite, quartz, iron compounds and various silicates. Very small amounts of (*i.e.* down to 0.1%) of sphalerite, copper/nickel/silver phases, ilmenite, rutile, sphene, and apatite were also sometimes found. The percentage of the non-identified phases was never >2% and was usually much less than this amount.

Table 8 gives the QEMSCAN results on the pot samples (*i.e.* MP1 to MP6). The presence of relatively high percentages (up to 13.2%) of lead compounds in three of the samples (MP2, MP4 and M6) was unexpected and is currently being further investigated. The major components for both MP1 and MP2 were various silicates (86.0 and 70.4%, respectively); for MP3, MP4 and MP5 it was calcite (54.5, 59.0 and 93.1%, respectively) and for MP6, it was gypsum/anhydrite (88.0%). Other significant amounts (*i.e.* >10%) of

**Table 6.** Information on six (Pharaonic) Egyptian kohl samples (and their containers)[g] plus initial analytical results

| Sample no. (approx. period[a]) [*Pot no.*] [Museum cat. No.] | Texture of contents | Colour of contents [*Pot colour*] [*Pot height*] | LV SEM[c] on contents (in decreasing order of wt. %[b]) [*LV SEM on the pots*[e]] | Probable main component(s) of the contents (kohl) [*Probable main component(s) of the pot*] |
|---|---|---|---|---|
| M1 (Middle Kingdom?) [*MP1*] [5/1946.771] | Powder | White/V. light brown [*Grey-Brown*] [*3.1 cm*] | Pb, O, C, Cl, Zn (Si, Ca, Fe, Al) [*Not done*[f]] | Lead carbonate/chloride/hydroxide [*Travertine?*[a] *(a form of* $CaCO_3$*)*] |
| M2 (Middle Kingdom?) [*MP2*] [5/1946.769] | Powder | White/V. light brown [*Grey-White*] [*3.8 cm*] | Pb, O, C, Fe, Si, Al (Na, Mg Ca) [*Not done*[f]] | Lead carbonate/hydroxide [*Travertine?*[a]] |
| M3 (New Kingdom?) [*MP3*] [5/1946.772] | Powder | Dark brown/Black [*Red-Brown*] [*5.9 cm*] | O, C, Ca, Si, Fe, Al, K, Mo (Cl, Na, Mg, Ti, Pb) [*Not done*[f]] | None (pot thought to be empty) [*"Egyptian ceramic"?*[a] *(sometimes "Nile mud")*] |
| M4 (Middle Kingdom?) [*MP4*] [357/1974/5] | Powder | Black [*Light Brown*] [*6.5 cm*] | C, O, Pb, S, Ca, Fe, Zn, Cl (Si, K, Al) [*O, C, Ca, Cu, Si (S, Mg)*] | Lead sulphide [*"Egyptian alabaster"*[a] *(that is Calcite)*] |

| | | | | |
|---|---|---|---|---|
| M5<br>(Middle Kingdom?)<br>[*MP5*]<br>[64 1919] | Powder | Dark brown/Black<br>[*Light Brown*]<br>[*4.5 cm*] | O, C, Pb, Ca, S, Cl, Al<br>(Si, Fe, Cu, Na)<br>[*O, Ca, C (Si, Mg)*] | Lead sulphide<br>[*"Egyptian alabaster"*[a]] |
| M6<br>(Middle Kingdom)<br>[*MP6*]<br>["Abydog 1922"][d] | Powder | White/Light brown<br>[*Blue-White*]<br>[*4.2 cm*] | Pb, C, O, Cl, S, Fe, Cu, Zn<br>(Ca, K, Si, Na, Al)<br>[*O, Ca, S, Si, Fe (K, Na, Al, Mg, C)*] | Lead carbonate/chloride/hydroxide<br>[*"Blue marble"*[a] *(that is Anhydrite)*] |

[a] Done by a local archaeologist (Dr. R.G. Morkot), on the basis of pot shape, size and colour, etc. (a "?" indicates that the approximate period/composition given is tentative).

[b] The elements in brackets are at less than 1% level each.

[c] For details of this technique see text.

[d] This pot is thought to have come from the excavation of Abydos in 1922 by W.M.F. Petrie.

[e] Done from *external* "scrapings" from each pot (pot data is given in *italics* in [...]).

[f] Insufficient material to do LV SEM.

[g] From the Royal Albert Memorial Museum (Exeter, UK).

compounds present were (apart from the lead compounds already mentioned above): gypsum/anhydrite (MP3 and MP4) and various silicates (MP3).

## 4. DISCUSSION

### 4.1. Comparison of past and present origins/compositions

The origins of the modern-day Egyptian kohl samples analysed are shown in Fig. 1 and it can be seen that almost three-quarters of them originate in Egypt. However, for the ancient samples analysed by others (see Tables 1 and 2) and by ourselves (see Tables 6–8), we can only speculate on their origins. It is known that almost all of the main components listed in Table 1 were available within Egypt itself; the one exception being stibnite, which probably came from Asia Minor (Turkey), certain of the Greek islands or possibly Arabia. It is also known that some of the eye-paints were imported from Naharin (in western Asia) and from Punt (Eritrea/Somalia) – the latter presumably being only a staging post for material that originated in Arabia. Both galena and malachite still occur in various parts of Arabia [3,29].

Of the 18 Cairo-purchased modern-day samples, 6 were found to be lead-based (*i.e.* 33%). If the 9 samples purchased in Luxor and Aswan are included, then this percentage falls to 22% (Fig. 2). This compares very favourably with the value of 63% for the lead-based samples in antiquity (Fig. 3; based solely on the data given in Table 1). Unfortunately the Louvre-based authors have so far *only* published analytical data on lead-based samples (for instance, given in Table 2) and thus their data were not included in the preparation of Fig. 3. Also, of the six Pharaonic kohl pots studied by us, high percentages (>75%) of lead phases were found in five of the pots (one was assumed to be empty of kohl as essentially no lead phases were found) (see Tables 6–8). The other phases listed in Table 7 are either contamination from the environment (*e.g.* quartz from ubiquitous sand), from the original lead ores (*e.g.* the silver phases) or from the pot itself (*e.g.* the various calcium compounds found). Also, some degree of mineralogical alteration is likely to have occurred as the pots have been open for an unknown period of time. However, regarding the lead

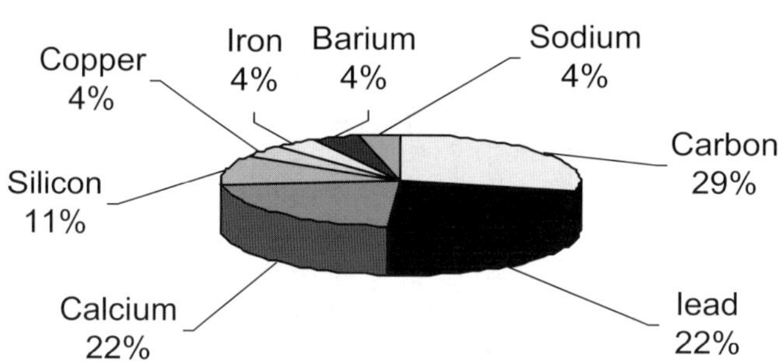


**Fig. 2.** Distribution of the main element of the major phase in the modern-day Egyptian kohl samples studied.

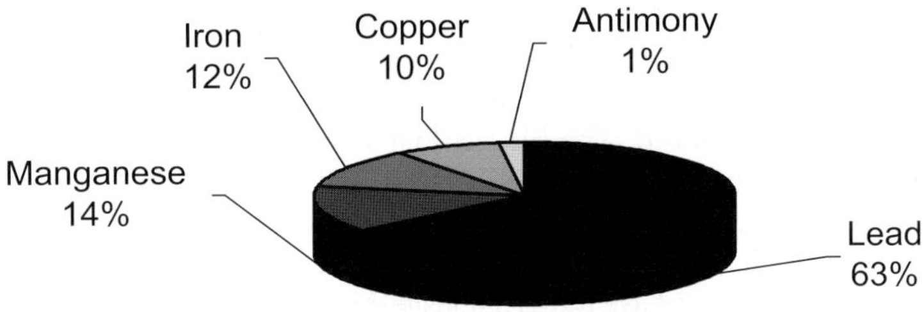


**Fig. 3.** Distribution of the main element of the major phase in ancient Egyptian kohl samples (Ref. [3]; Table 1 only).

phases, as previously stated we cannot distinguish definitely between the various lead phases that are present in these samples. Further work on this is in progress and will appear in a later publication. The other major differences between past and present compositions are: the presence of manganese and (once) antimony compounds only in ancient samples; the absence of any analysed ancient samples that consist mainly of carbon; and the presence of barium, silicon and calcium compounds, as the major components, in only modern-day samples. Also, in the past, the colour of the sample was due to the main component, though various "whitening" materials were added sometimes (see next section), as compared to the present-day, when the colour is given by a small percentage (1–2%) of a (probably synthetic and organic) colourant added to the (white, inexpensive and readily available) main component (such as calcite or talc) (see for eight of the nine samples in Table 4 and for three of the 18 samples in Table 3).

As already mentioned, only one "old" (analysed) sample, from the nineteenth dynasty, has been found to be an antimony compound (stibnite, $Sb_2S_3$) (see Table 1). The reason for the occasional statement that antimony/antimony (tri)sulphide/stibnite was used as an eye cosmetic in ancient Egypt (*e.g.* Refs. [30,31]) is mainly one of philology. The ancient Egyptian word for eye-paint in general and the black form in particular was "msdmt" (mesdemet) and it became "cthm" (stem) in Coptic, then "stimmi" in Greek and finally "stibium" in (Roman) Latin. This last word was later used for the element antimony, and stibnite for its sulphide ore. Also, as already mentioned, in their massive states the two ores stibnite and galena look very similar. Stibnite is rarer than the more common (and cheaper) galena and so the temptation to replace the former with the latter would have been an "economic incentive" from the earliest times.

Also, one of the authors can personally testify to being offered "ithmid"/"ethmid"/ "athmid" (that is the eye cosmetic of Islam, which has generally been assumed to be the ore stibnite) in various souks of modern-day Arabia and subsequently finding them all to be the ore galena. To our knowledge no early-Islam sample, from a reputed museum and of known provenance, has yet been chemically analysed.

As to the speculated usage of lapis lazuli as a blue eye cosmetic by Cleopatra VII (see Introduction), there is presently no analytical published data showing that lapis lazuli was used as an eye cosmetic in ancient Egypt. Funerary cosmetic items from Pharaonic Egypt, that have been subjected to detailed chemical analytical study and subsequently published in peer reviewed and abstracted journals, do not currently include any definitive blue material. Chrysocolla (a hydrated copper silicate) has been found once, but its colour can vary from blue-green to green. However, a modern-day study of a North African recipe for blue eye-shadow showed that by subjecting natural galena to "heat treatment", a blue-appearing material is produced. The blue colour is in fact an optical interference effect resulting from the formation of layers of anglesite ($PbSO_4$) and then of lanarkite ($PbO.PbSO_4$) on the original galena [32]. It is possible that these materials, and their effect, were produced accidentally in ancient Egypt. However, the recent analyses of many eye cosmetics stored in the Louvre have found anglesite, but not lanarkite, to be present sometimes (*e.g.* see Table 2).

## 4.2. Toxicology of lead

Lead compounds are toxic by ingestion, inhalation and by skin exposure. Children are more susceptible than adults to lead intoxication. The toxic effects of lead form a continuum from clinical or overt effects to more subtle ones [33]. The critical effects in infants and children involve the nervous system. Blood lead levels once thought to be safe have been shown to be associated with intelligence quotient deficits, behavioural disorders, slowed growth and impaired hearing [34,35]. Blood lead levels in children that are greater than 10 μg/dl are now considered abnormal [36], and recently it has been shown that significant intellectual impairment occurs in young children who have blood lead levels below 10 μg/dl [37]. Severe lead poisoning, resulting in encephalopathy, can result when the blood lead levels are greater than 70 μg/dl. A recent report has demonstrated that young infants exposed to lower levels of lead following the use of traditional medicines can also present with encephalopathy [38]. Reported cases of acute encephalopathy in infants that are directly linked to excessive usage of a lead-based kohl are now fewer than several decades ago, but unfortunately still do occur [39].

Frequently, mothers apply kohl to infants and children as a traditional measure to beautify and protect the child from the "evil eye". Lead-based kohls can be easily ingested by these infants who may wipe their eyes and face and subsequently lick their fingers; earlier (animal) studies [40] have shown that transcorneal transport is not a significant contributory mechanism for absorption of lead from lead-based eye cosmetics. It is worth noting that adults absorb 5–15% of ingested lead while children can absorb as much as 41% of ingested lead.

Solubility studies have shown that the particle size of the ground galena is directly related to its rate of dissolution (*i.e.* conversion to the more soluble, and hence more readily absorbed, chloride forms) in gastric fluid. An increase in the rate of dissolution, by a factor of approximately two, was found for galena of mean particle size 30 μm as compared to galena with a mean particle size of 100 μm [40]. As a result of this effect, larger sized particles of galena could well pass through the GI (gastro-intestinal) tract before it is converted to a more readily absorbed form. This particle-size effect could well explain the varying degrees to which galena has been reported to be absorbed in the gut. As the galena powder is ground it loses its initial high iridescence to become progressively more matt in appearance, becoming totally matt at a mean particle size of about ≤10 μm. In this study, the two galena-based Indian-made kohl powders (sample nos. 1 and 2; with sample no.1

found to have an estimated average particle size of 5 μm) are totally matt in appearance; the two galena-based samples made in Egypt (sample nos. 5 and 6; with sample no. 6 found to have an estimated average particle size of 11 μm) are mostly matt; and the two hand-ground samples (sample nos. 3 and 4; with sample no. 3 found to have an estimated average particle size of 69 μm) are highly iridescent. Thus if these six galena-based samples were equally ingested there would be a range in the rates of their dissolution in the stomach's gastric fluids, with the two Indian-made samples having the highest rates of dissolution, and hence absorption.

As the present, so the past. A range of powder appearances (from wholly matt to wholly highly iridescent), and hence a range of particle sizes, have been found for the galena in some of the ancient Egyptian kohl samples in the Louvre. Using data from several techniques (SEM and Transmission electron microscopy (TEM) images for particle sizes, and peak profile analysis of synchrotron XRD data for crystallite sizes), it has recently been suggested [14] that there were four or five "manufacturing procedures" used by the ancient Egyptians for making kohl. These are: (1) gentle crushing, (2) gentle crushing and sieving, (3) crushing, (4) crushing and sieving and perhaps (5) crushing and heating (to 200–300°C). Kohls made by "procedures" (1) and (2) (and (5), if used) are iridescent and those from the two other "procedures" are mainly matt. Possibly this was done in response to a demand for differing styles of eye cosmetic. These variously made powders would then perhaps have been mixed with a naturally occurring white dilutant (such as cerussite) to give shades of grey cosmetics and/or mixed with one or both of the "made" lead compounds (phosgenite and laurionite) to give eye salves or possibly face cosmetics. These two "made" compounds have been found to have smaller particle sizes (down to 1 μm).

More than 90% of lead in blood resides in the red blood cells. The total body burden of lead can be divided into two kinetic pools, which have different rates of turnover. The largest pool is in the skeleton, which has a very slow turnover (a half life of more than 20 years) [36]. The other pool is in the soft tissue, where it is much more labile. Lead in the trabecular bone is more labile than in the cortical bone, and trabecular bone has a shorter turnover time. Lead in bone may contribute up to 50% of blood lead. During pregnancy and lactation, mobilisation of lead from maternal bone is a cause for concern. Strong correlations between maternal and umbilical cord blood lead levels demonstrate that lead is transferred from the mother to the foetus [41,42]. Cumulative effects of low levels of lead exposure *in utero* and after birth can have similar detrimental effects. An increase in maternal-blood lead level may contribute to a reduction in gestation period and low birthweight. The foetal brain may also be particularly sensitive to the toxic effects of lead because of the immaturity of the blood–brain barrier.

In an adult population the most critical adverse effect of lead is probably hypertension. Other toxic effects of concern are peripheral neuropathy, lead-induced anaemia and lead nephropathy.

In view of the above mentioned toxic effects and the still widespread use of kohls in the present-day Middle East, it follows that children who have a lead-based kohl regularly applied to them are at risk of serious and fatal toxicities of the nervous system and also to more subtle, subclinical, long-term effects.

### 4.3. Written information on container/packaging

In ancient (Pharaonic) Egypt, the funerary containers used for eye cosmetics and/or eye salve were made of materials such as: glass (New Kingdom onwards), wood, reed, bone, steatite (also known as soapstone, a massive form of talc), serpentine (a magnesium silicate), ivory and/or ebony, obsidian (a glassy volcanic rock), rock crystal (a form of quartz), alabaster (a fine-grained massive form of gypsum, that is hydrated calcium sulphate), "Egyptian alabaster" (which is in fact calcite, calcium carbonate), anhydrite (the anhydrous form of calcium sulphate, the so-called "blue marble" of ancient Egypt) or "Egyptian ceramic" (also known as "Nile mud"; mostly a mixture of silicates). Also, multiple containers (*i.e.* two, three or four joined containers) are known and it is assumed that these were for kohls to be used in different seasons and/or contained kohls of different colours. On some containers, both single and multiple, are written, in hieroglyphs, comments such as: "Genuine, very excellent kohl"; "Opens vision" (*i.e.* an eye salve/solution); "Repels blood" (*i.e.* checks bleeding) [1] and "Good for the sight"; "To cause tears"; "For daily use" [43].

The Pharaonic kohl pots studied by us (see Fig. 4) were originally thought to be composed of the materials listed in Table 6 (under *"Probable main component(s) of the*

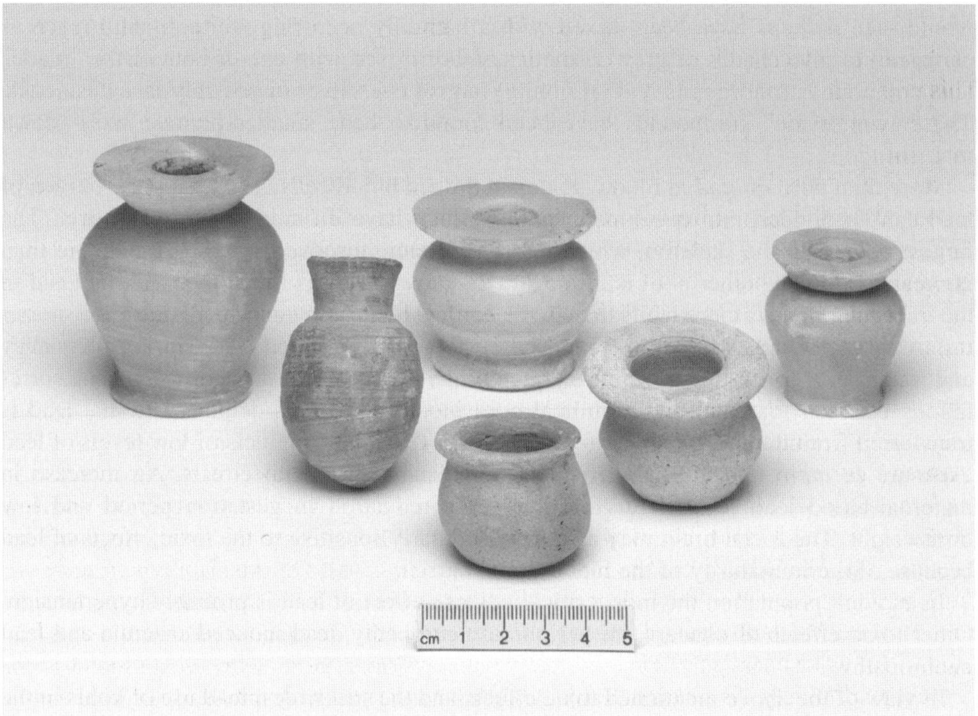


**Fig. 4.** The six Pharaonic kohl pots studied. Pot nos. (from L to R and back row first): MP4, MP5, MP6, MP3, MP1 and MP2 (see Tables 6, 7 and 8) (© Royal Albert Memorial Museum and Art Gallery, Exeter, UK).

*pot itself*"). The analytical results in Tables 6 and 8 show that these were correct for pots MP5 and MP6; *i.e.* "Egyptian Alabaster" (calcite) for MP5 (at 93.1%) and "blue marble" (anhydrite) for MP6 (at 88.0% for gypsum/anhydrite in Table 8 and anhydrite rather than gypsum from Table 6). However for MP1 and MP2, the expected travertine (a naturally occurring calcium carbonate deposit) was in fact found to be incorrect and their main components (at 86.0 and 70.4%, respectively) were "various silicates" (mostly calcium aluminium silicates). This indicates that these pots are probably made of "Egyptian ceramic" (see above). For MP4, a significant amount (59.0%) of calcite was found, but additionally significant amounts of gypsum/anhydrite (at 26.2%) were also found. Also, for pot MP3, high percentages of calcite (54.5%) and gypsum/anhydrite (26.2%) were found. If these pots had been made from "Egyptian ceramic", then significant amounts of silicates would have been expected. This indicates that the pots MP3 and MP4 were possibly made from a mixture of "Egyptian alabaster" and alabaster (that was then a naturally occurring sedimentary deposit).

As regards the contents of the containers, the hieroglyph for "msdmt" is sometimes seen and in one case the hieroglyph for "high quality", repeated three times, is placed before it [17]. As already stated, the word "msdmt" is used for eye-paint in general and the black form in particular. When this hieroglyph was observed on one sample of a nineteenth dynasty funerary deposit it was found to be galena (lead sulphide), whilst other samples from the same funerary deposit but with hieroglyphs saying "eye lotion to be dispersed, good for eyesight", were found to be mixtures of lead chloride and lead carbonate [16]. This provides evidence that the ancient Egyptians (here, New Kingdom) had a knowledge of "wet chemistry" and made these lead compounds with the definite intention of using them for therapeutic purposes.

Out of the 18 modern-day Cairo samples studied, 12 had a name written on the container and/or on the packaging. Of the 9 samples from Aswan and Luxor, only 5 had a name on the container. Table 5 lists the names as found on the labels/packaging. These names are usually in English, but in two cases no English name was found and the translation of the name (from Arabic) is given. Also given in this table, on a simple Y/N basis, is whether any information is given on the contents, on the medicinal effects of using the kohl and whether the sample was found to contain lead.

Regarding the data on the contents, only 2 samples (nos. 1 and 2) gave definite *quantitative* "contents formulas" on enclosed leaflets (as percentages). Sample no. 1 has "Asmad" given as 100% of its contents and this is often found to be lead sulphide (as here). Sample no. 2 gives the following contents data (with the percentages given in brackets): Asmad (70%); Bh. Kafoor (28.5%); Sadaf Softi (Mori) (0.70%); Ark Phudina (0.30%); Ph. Ph. Safed (0.25%) and Hab-El-Arus (0.25%). On translation/interpretation these substances are: lead sulphide, camphor, "pearl" (*i.e.* probably the mineral aragonite, or possibly calcite – both being forms of calcium carbonate); extract of mint; "white potassium?" (the exact chemical name is currently unknown) and "Bean of the bride-groom" (Java pepper) respectively. As given in Table 3, both these samples contain lead sulphide as the major phase and for sample no. 2 camphor is also found. Additionally, both these samples contain small amounts of anglesite and cerussite, with sample no. 2 also having a small amount of zincite. No other labelled sample gave any contents data on its container or on its packaging.

As regards the data on medicinal effects, it was found that only two samples (sample nos. 2 and 9) gave this information – in an enclosed leaflet in each case. Both samples were made in Bombay (India), but not by the same company. For sample no. 9 (Hind Ka Noor eye liner), the data was written in colloquial Arabic and (after translation) it stated that "it was good for": "Reducing cold", "Eye ache", "All diseases of the eye", "Heat in the eye", and "Improves eye-sight and strengthens vision". It also stated that "Used on a daily or weekly basis it will protect you always against diseases of the eye" and "can be used by adults or children". This particular kohl has been seen before, in the souks of: Amman (Jordan) (unpublished data), Abu Dhabi city (Ref. [44]; sample no. 11), Bahrain and Oman (Ref. [45]; sample nos. 27 and 28 for Oman and no. 31 for Bahrain). The major phase is always amorphous carbon and the minor phase(s) are one or more of the following: dolomite ($(Mg,Ca)(CO_3)_2$), graphite, quartz and talc. All these substances are unlikely to give rise to adverse medical conditions when used externally on the eye; however, if used internally (*e.g.* on the conjunctiva surface) then abrasions could well occur, especially from the harder substances present (*i.e.* quartz and, to a lesser extent, dolomite and graphite). Also, the larger and more irregular the compound's particles, then greater will be the potential for eye abrasions.

Sample no. 2 (Khojati Surma No. 9) has the written information (in English): "It is cooler than Khojati Surma No. 13 and of better quality. Its regular use keeps the eyes clean, healthy and reduces the adverse effects caused by heat". Khojati Surma No. 13 was one of the samples purchased in Abu Dhabi city souk [44] and was found to contain calcium carbonate as the major phase and with camphor, kaolinite, iron silicate hydrate and graphite as the minor phases. As stated before, these substances are unlikely to give toxicity; however, this is not the case for our sample no. 2, where the major phase is a lead compound (galena). Thus, the use of this kohl could give rise to lead toxicity and its quoted medicinal effects are questionable at best and dangerous at worst.

Additionally, found only in the enclosed leaflet of sample no. 1 (Khojati Surma Sada; also sample no. 1 in our Abu Dhabi city study), is the following statement (in English): "This is the purest form of Surma scientifically ground in different extracts. As there is no addition of any other medications, it does not make the eye water *and may be used for children below the age of 8 years*" (our addition in *italics*). This is identical to the wording found along with the sample purchased in Abu Dhabi and as then, we can only repeat our view that such advice, especially with regard to young children, is both dangerous and outrageous as the major phase is (as before) lead sulphide *and* is of such a small particle size that it will be readily absorbed by the body's gastric juices.

Samples nos. 1 and 2 are also the only ones to have written data on how to apply the kohl to the eye. Both contain a plastic applicator rod in the sample box; ten other samples (nos. 5, 6, 7, 10, 12, 13, A2, A3, A4 and L5) have an applicator rod actually in the container with the kohl powder. The written data for samples 1 and 2 states (in English): "Apply with a clean and dry salai (applicator) a minimum quantity of surma in the morning and half an hour before going to bed". Sometimes, in present-day Egypt, the powder is applied as is (*i.e.* dry), sometimes after mixing with water to make a paste and sometimes after mixing with "an oil" (often olive oil). The Egyptian Bedouin are known to use the juice of chopped onions on the applicator before placing it in a kohl powder and then on/in/around the eye. The resulting "washing of the eye" is regarded as a beneficial side-effect of using the kohl. It is also reported that in modern-day Upper Egypt (black) kohl, which has had both lemon juice and onion juice added to it *via* the use of an applicator, is placed in/on/around a new-born baby's eyes once a week for the first 40 days of its life. This is done for two reasons; to kill germs (the juices) and to take away the "evil eye" (the kohl) (Hardy; personal communications, 2001).

Of the 17 labelled samples, four were found to contain lead. Two have already been discussed (sample nos. 1 and 2). The other two were sample nos. 5 and 6; both of these samples give no mention of containing lead on the label, but in addition to the name (CHOL NORHAN) there was a "cold" (Arabic) symbol on the label of sample no. 5 and a "hot" (Arabic) symbol on the label of sample no. 6. A "hot" kohl sample is often used as an eye medicine and so is sometimes placed *inside* the eye, whilst a "cold" kohl sample is used solely for beautification and is often only used on the outside of the eye. In Oman, a kohl used as a medicine (*i.e.* "hot") often contained, or was supposed to contain, an "active ingredient" of a lead or antimony compound [21]. Here, *both* "hot" and "cold" labelled kohl samples have lead sulphide as the major phase; but the third CHOL NORHAN sample (no. 7) purchased, which had no such "hot"/"cold" label, but was designated "cold" by the shopkeeper, consists solely of amorphous carbon.

## 5. CONCLUSIONS

As has been stated before, lead is of no known biological value, is not an essential element and when present in the blood can give rise to toxicity. In this study, lead was found to be present in almost one-quarter of the Egyptian modern-day samples studied. Six samples were based on galena, six on amorphous carbon, three on silicon/silicon-based compound (talc), six on calcite, one on cuprite, one on goethite, one on barite, one on halite and two on unknown (but assumed to be carbon-based) amorphous compounds. If the other "heavy metals" (copper and iron, *i.e.* those metals with a density $\geq 5$ gms/c.c.) found as a major phase element in these samples are included, then the above percentage rises from 22% (only lead-containing) to 30% (copper, iron and lead-containing). In the (Pharaonic) past a higher percentage (63%, using the Table 1 data only) of the samples were lead-based. This was confirmed by our findings; five of the six Pharaonic kohls studied by us had lead-based contents (one was empty). Two were black in colour and so probably contained lead sulphide, whilst the other three were white in colour and probably consisted of a mixture of various natural/"made" lead compounds/minerals (such as cerussite, laurionite, phosgenite, hydro-cerussite and anglesite). It appears unlikely that the ancient Egyptians understood lead's toxic nature as they deliberately used the black eye-paint, and other "made" lead compounds, in the treatment of eye diseases/infections.

Now and then, young children are especially vulnerable to lead toxicity; with the level of lead in blood that can cause long-term damage being continuously revised downwards. It now appears that there may be no lower limit for adverse effects from lead exposure to occur and that once the impairments have occurred they may well be both persistent and irreversible. Thus it is essential that the initiatives started by the Egyptian Government/USAID/CDC be continued and that there should continue to be a reduction in the availability of lead-containing eye cosmetics in modern-day Egypt.

**Table 7.** QEMSCAN results (wt. %s)[a] on the contents of six (Pharaonic) Egyptian kohl pots

| Mineral | M1 (%) | M2 (%) | M3 (%) | M4 (%) | M5 (%) | M6 (%) |
|---|---|---|---|---|---|---|
| Lead phases[b] | 91.3<br>(94.0) | 87.4<br>(90.9) | 0.2<br>(0.1) | 79.5<br>(81.4) | 78.4<br>(80.4) | 91.5<br>(92.3) |
| Calcite ($CaCO_3$) | 1.4<br>(0.7) | 1.8<br>(0.8) | 27.9<br>(18.2) | 6.5<br>(4.7) | 11.0<br>(8.5) | 0.3<br>(0.2) |
| Gypsum/Anhydrite[b] ($CaSO_4.2H_2O$/ $CaSO_4$) | 0.3<br>(0.4) | 3.3<br>(4.3) | 0.9<br>(1.9) | 3.8<br>(5.3) | 4.1<br>(5.4) | 0.4<br>(0.5) |
| Quartz ($SiO_2$) | 3.7<br>(1.1) | 1.0<br>(1.0) | 17.7<br>(12.3) | 3.7<br>(2.6) | 0.9<br>(0.7) | 1.6<br>(0.8) |
| Iron phases | 1.3<br>(1.0) | 3.7<br>(1.5) | 1.2<br>(0.2) | 3.0<br>(2.4) | 3.1<br>(3.0) | 2.1<br>(1.8) |
| Sphalerite (ZnS) | 0.6<br>(1.3) | 0.0<br>(0.0) | 0.0<br>(0.0) | 0.5<br>(0.8) | 0.5<br>(0.6) | 0.4<br>(0.4) |
| Copper phases | 0.0<br>(0.0) | 0.0<br>(0.0) | 0.0<br>(0.0) | 0.0<br>(0.0) | 0.0<br>(0.0) | 0.7<br>(1.0) |
| Silver phases | 0.0<br>(0.0) | 0.2<br>(0.0) | 0.0<br>(0.0) | 0.0<br>(0.5) | 0.0<br>(0.0) | 0.0<br>(1.6) |
| Nickel phases | 0.0<br>(0.0) | 0.0<br>(0.0) | 0.0<br>(0.0) | 0.2<br>(0.1) | 0.6<br>(0.5) | 0.0<br>(0.0) |
| Ilmenite/Rutile ($FeTiO_3$/$TiO_2$) | 0.0/0.0<br>(0.0/0.0) | 0.0/0.0<br>(0.0/0.1) | 0.5/0.2<br>(0.7/1.0) | 0.0/0.1<br>(0.2/0.1) | 0.2/0.0<br>(0.2/0.0) | 0.0/0.1<br>(0.1/0.0) |
| Sphene/Apatite ($MgAl_2O_4$/ $Ca_5(PO_4)_3$ (F, Cl, OH)) | 0.0/0.0<br>(0.0/0.0) | 0.0/0.0<br>(0.0/0.0) | 0.1/0.9<br>(0.4/0.6) | 0.0/0.1<br>(0.0/0.0) | 0.0/0.0<br>(0.0/0.0) | 0.0/0.0<br>(0.0/0.0) |
| Various Silicates | 1.4<br>(1.5) | 1.6<br>(1.1) | 50.1<br>(64.6) | 2.2<br>(1.8) | 0.9<br>(0.7) | 0.9<br>(0.7) |
| Non-identified phases | 0.0<br>(0.0) | 1.0<br>(0.3) | 0.3<br>(0.0) | 0.4<br>(0.1) | 0.3<br>(0.0) | 2.0<br>(0.6) |
| Number of particles analysed | 10 050<br>(5019) | 10 012<br>(661) | 10 225<br>(2706) | 10 132<br>(5012) | 10 106<br>(5107) | 10 012<br>(4027) |

[a] Two size fractions of the kohls were analysed. Particles between 1 and 25 μm; and the values in brackets are for particles between 25 and 200 μm (see Experimental section for details).

[b] In this technique, energy dispersive spectrometry cannot currently distinguish between the various lead phases present, and also between Gypsum and Anhydrite.

**Table 8.** QEMSCAN results (wt. %s) on the composition of six Pharaonic kohl pots

| Mineral | MP1 (%) | MP2 (%) | MP3 (%) | MP4 (%) | MP5 (%) | MP6 (%) |
|---|---|---|---|---|---|---|
| Lead phases | 0.5 | 13.0 | 0.5 | 13.2 | 1.9 | 8.3 |
| Calcite | 0.8 | 1.5 | 54.5 | 59.0 | 93.1 | 0.2 |
| Gypsum/Anhydrite | 1.3 | 3.5 | 26.2 | 21.5 | 1.3 | 88.0 |
| Quartz | 7.3 | 9.1 | 1.9 | 1.0 | 0.4 | 0.5 |
| Iron phases | 0.3 | 0.4 | 0.4 | 0.6 | 0.0 | 0.1 |
| Sphalerite | 0.0 | 0.0 | 0.0 | 0.2 | 0.0 | 0.1 |
| Copper phases | 0.0 | 0.0 | 0.0 | 0.8 | 0.0 | 0.0 |
| Ilmenite/Rutile | 0.1/0.1 | 0.0/0.2 | 0.2/0.0 | 0.6/0.0 | 0.0/0.0 | 0.0/0.0 |
| Sphene/Apatite | 0.1/0.1 | 0.1/0.1 | 0.0/0.2 | 0.0/0.0 | 0.0/0.0 | 0.0/0.0 |
| Nickel phases[a] | 3.1 | 1.4 | 0.0 | 1.2 | 2.8 | 0.4 |
| Various Silicates | 86.0 | 70.4 | 16.1 | 1.8 | 0.5 | 2.3 |
| Non-identified phases | 0.3 | 0.3 | 0.0 | 0.1 | 0.0 | 0.1 |
| Number of particles analysed | 5065 | 5053 | 5019 | 4226 | 5044 | 5074 |

[a] These percentages are a summation for Nickel sulphides and elemental Nickel (which is assumed to have come from the sampling spatula).

## ACKNOWLEDGEMENTS

We would like to thank the following people for their help in the course of this study: Mr. P. Auchterlonie (Librarian for Middle East Studies, Exeter University, UK) and Dr. K. A. Mahdi (Institute of Arab and Islamic Studies, Exeter University, UK) for their help in translating colloquial Arabic and to Prof. P. Pattie, Ms. Marian M. Azmy and Mrs. Amany Wilson (faculty member, student and staff, respectively at the American University of Cairo, Egypt) for their help, during a visit to Egypt by one of us (ADH), in obtaining both samples and ethnographic data. Also, we would like to thank the staff of the Chemical and Materials Analysis Unit (University of Newcastle, UK) for the experimental LV SEM work mentioned here.

We would like to thank the various staff members at the Royal Albert Memorial Museum (Exeter, UK) for access to the Pharaonic pots and also the "scrapings" from them. Also to a local archaeologist (Dr. R.G. Morkot; Department of Lifelong Learning, Exeter University, Exeter, UK) for giving us his opinion as to their age, provenance and composition. A much shorter paper (c. 3000 words), based solely on the present-day Cairo "kohls" analytical results and without any of the detailed archaeology/history/ethnographic data given here, was previously published in the International Journal of Environmental Health Research [46].

## REFERENCES


[1] L. Manniche, W. Forman (photographer), *Sacred luxuries: Fragrance, Aromatherapy and Cosmetics in Ancient Egypt*, Opus Publishing Ltd., London, pp. 124–142, 1999.

[2] R. Friedman, Excavating Hierakonpolis: Who let the Elephants in? Archaeology's Interactive Dig, www.archaeology.org/interactive/hierakonpolis/field/elephant1.html, 2003.

[3] A. Lucas, J.R. Harris, *Ancient Egyptian Materials and Industries*, fourth ed., Dover Publications, New York, pp. 80–84, 99, 113, 195–199, 210, 243–244, 262, 340–344, 398–401, 1962.

[4] R.J. Forbes, *Studies in Ancient Technology*, vol. III, third ed., E.J. Brill, Leiden, The Netherlands, pp. 17–20, 1993.

[5] B. Aston, J. Harrell, I. Shaw, *Stone*, P.T. Nicholson, I. Shaw, (Eds.), in: *Ancient Egyptian Materials and Technology*, Cambridge University Press, Cambridge, UK, pp. 44, 107, 148, 2000.

[6] J.F. Nunn, *Ancient Egyptian Medicine*, British Museum Press, London, pp. 138, 146, 147, 198, 199, 201, 1996.

[7] W.R. Dawson, *Magician and Leech: a Study in the Beginnings of Medicine with Special Reference to Ancient Egypt*, Methuen and Co. Ltd., London, UK, pp. 109–125, 1929.

[8] H.E. Sigerist, *A History of Medicine, Volume I: Primitive and Archaic Medicine*, Oxford University Press, UK, pp. 216–373, 1951 (third printing, 1977).

[9] R. Porter, *The Greatest Benefit to Mankind: a Medical History of Humanity from Antiquity to the Present,* Harper Collins, London, UK, pp. 47–50, 1997.

[10] S.R. Andersen, The eye and its diseases in ancient Egypt, *Acta Ophthalmologica Scandinavica*, **75** (1997) 3, 338–344.

[11] M. Angeloglou, *A History of Make-up*, Studio Vista Limited, London, UK, p. 22, 1970.

[12] P. Bancroft, *Gem and Crystal Treasures*, Western Enterprises/Mineralogical Record, California, USA, 1984.

[13] J. Barthoux, Les fards, pommades et couleurs dans l'antiquite. Congres Int. de Geog., Le Caire, Avril 1925, IV (1926), pp. 251–267.

[14] T. Ungar, P. Martinetto, G. Ribarik, E. Dooryhee, Ph. Walter, M. Anne, Revealing the powdering methods of black makeup in Ancient Egypt by fitting microstructure based Fourier coefficients to the whole X-ray diffraction profiles of galena, *Journal of Applied Physics*, **91** (2002) 4, 2455–2465.

[15] E. Dooryhee, Advances in Art and Archaeology using X-ray synchrotron radiation, in: G. Tsoucaris, J. Lipkowski (Eds.), *Molecular and Structural Archaeology: Cosmetic and Therapeutic Chemicals*, (Proceedings of the NATO Advanced Research workshop, Erice, Sicily, May 2002), Kluwer Academic Publishers, Dordrecht, The Netherlands, pp. 201–209, 2003.

[16] Ph. Walter, P. Martinetto, G. Tsoucaris, R. Breniaux, M.A. Lefebvre, G. Richard, J. Talabot, E. Dooryhee, Making make-up in Ancient Egypt, *Nature*, **397** (1999), 483–484.

[17] P. Martinetto, M. Anne, E. Dooryhee, G. Tsoucaris, Ph. Walter, A synchrotron X-ray diffraction study of Egyptian cosmetics, in: D.C. Creagh, D.A. Bradley, (Eds.), *Radiation in Art and Archeometry*, Elsevier Science BV, pp. 297–316, 2000.

[18] R.H. Brill, I.L. Barnes, B. Adams, Lead isotopes in some ancient Egyptian objects, A. Bishay (Ed.), in: *Recent Advances in Science and Technology of Materials*, vol. 3, Plenum Press, New York, USA, pp. 9–27, 1974.

[19] Z.A. Stos-Gale, N.H. Gale, Sources of galena, lead and silver in Predynastic Egypt, *Revue d'Archeometrie*, **5** (1981), 285–295.

[20] A.A. Hassan, F.A. Hassan, Source of galena in predynastic Egypt at Nagada, *Archaeometry*, **23** (1981) 1, 77–82.

[21] A.G. Miller, M. Morris, S. Stuart-Smith (illustrator), *Plants of Dhofar*, The Office of the Advisor for Conservation of the Environment, Diwan of Royal Court, Sultanate of Oman, p. 20, 1988.

[22] E.W. Lane, *Manners and Customs of the Modern Egyptians*, Everyman's Library edition, J.M. Dent and Sons Ltd., London, pp. 37, 38, 1908.

[23] R. Chappell, P. Billig, E. Brantly, S. Ault, H.S. Ezzeldin, *Lead Exposure Abatement Plan for Egypt: Results of Environmental Sampling for Lead*, Environmental Health Project, Activity Report No. 32 (prepared for the USAID Mission to Egypt under EHP Activity No. 256 – CC), pp. 44, 45, 69, 89, 90, 92, 1997.

[24] ICDD (International Centre for Diffraction Data), 12 Campus Boulevard, Newtown Square, PA 19073–3273, USA, 2000.

[25] P. Gottlieb, G. Wilkie, D. Sutherland, E. Ho-Tun, S. Suthers, K. Perera, B. Jenkins, S. Spencer, A.R. Butcher, J. Rayner, Using quantitative electron microscopy for process mineral applications, *Journal of Mineralogy*, **52** (2000), 24–25.
[26] G.S. Camm, A.R. Butcher, D. Pirrie, P.K. Hughes, H.J. Glass, Secondary mineral phases associated with a historic arsenic calciner identified using automated scanning electron microscopy; a pilot study from Cornwall, UK, *Minerals Engineering*, **16** (2003) (11, 1), 1269–1277.
[27] D. Pirrie, A.R. Butcher, M.R. Power, P. Gottlieb, G.L. Miller, Rapid quantitative mineral and phase analysis using automated scanning electron microscopy (QEMSCAN); potential applications in forensic geoscience, in: K. Pye, D.J. Croft (Eds.), *Forensic Geoscience: Principles, Techniques and Applications*, Geological Society, London, Special Publication 232, pp. 123–136, 2004.
[28] A.D. Hardy, *Inorganic Ethnic Remedies of Oman and the GCC States*, Umfreville Publishing, Argyll, UK, p. 58, 1999.
[29] M.A. Haddadin, C. Zachariah, *Minerals of Oman*, Ministry of Petroleum and Minerals, Sultanate of Oman, pp. 18, 32, 1987.
[30] Z. Goffer, *Archaeological Chemistry*, John Wiley and Sons, New York, USA, p. 208, 1980
[31] B. Selinger, *Chemistry in the Marketplace*, fourth ed., Harcourt Brace and Company, Florida, USA, p. 97, 1994.
[32] B. Ponsot, J. Salomon, Ph. Walter, RBS study of galena thermal oxidation in air with a 6-MeV $^{16}O^{3+}$ ion beam, *Nuclear Instruments and Methods in Physics Research B*, (1998) 136–138, 1074–1079.
[33] R.A. Goyer, Lead toxicity: from overt to subclinical to subtle health effects, *Environment Health Perspective*, **86** (1990), 177–181.
[34] H.L. Needleman, A. Schell, D. Bellinger, E.N. Alfred, The long term effects of exposure to low doses of lead in childhood: an eleven year follow-up report, *New England Journal of Medicine*, **332** (1990), 83–88.
[35] J. Schwartz, D. Otto, Lead and minor hearing impairment, *Archives Environmental Health*, **46** (1991), 300–06.
[36] C.D. Klaasen, *Casarett and Doull's Toxicology, the Basic Science of Poisons*, fifth ed., McGraw-Hill, New York, p. 704, 1996.
[37] R.L. Canfield, C.R. Henderson, D.A. Cory-Slechta, C. Cox, T.A. Jusko, B.P. Lanphear, Intellectual impairment in children with blood lead concentrations below 10 μg per decliter, *New England Journal of Medicine*, **348** (2003) 16, 1517–1526.
[38] A. Al-Khayat, N.S. Menon, M.R. Alidina, Acute lead encephalopathy in early infancy – clinical presentation and outcome, *Annals of Tropical Paediatrics*, **17** (1997), 39–44.
[39] H. Narchi, Radiological Case of the Month, *Achives of Pediatric and Adolescent Medicine*, **154** (2000), 83–84.
[40] M.A. Healy, P.G. Harrison, M. Aslam, S.S. Davis, C.G. Wilson, Lead sulphide and traditional preparations: routes for ingestion, and solubility and reactions in gastric fluid, *Journal of Clinical and Hospital Pharmacy*, **7** (1982), 169–173.
[41] R.A. Goyer, Transplacental transport of lead, *Environment Health Perspective*, **89** (1990), 101–108.
[42] C. Gardella, Lead exposure in pregnancy: a review of the literature and argument for routine prenatal screening, *Obstetrical and Gynaecological Survey*, **56** (2001) 4, 231–238.
[43] M. Dayagi-Mendels, *Perfumes and Cosmetics in the Ancient World*, The Israel Museum, Jerusalem, p. 44, 1989.
[44] A.D. Hardy, H.H. Sutherland, R. Vaishnav, A study of the composition of some eye cosmetics (kohls) used in the United Arab Emirates, *Journal of Ethnopharmacology*, **80** (2002), 137–145.
[45] A.D. Hardy, R. Vaishnav, S.S.Z. Al-Kharusi, H.H. Sutherland, M.A. Worthing, Composition of eye cosmetics (kohls) used in Oman, *Journal of Ethnopharmacology*, **60** (1998), 223–234.
[46] A.D. Hardy, R.I. Walton, R. Vaishnav, Composition of eye cosmetics (kohls) used in Cairo, *International Journal of Environmental Health Research*, **14** (2004) 1, 83–91.

# Author Index